The two volumes that comprise this work provide a comprehensive guide and source book on the marine use of composite materials.

The first volume, *Fundamental Aspects*, provides a rigorous development of theory. Areas covered include materials science, environmental aspects, production technology, structural analysis, finite-element methods, materials failure mechanisms and the role of standard test procedures. An appendix gives tables of the mechanical properties of common polymeric composites and laminates in marine use.

The second volume, *Practical Considerations*, examines how the theory can be used in the design and construction of marine structures, including ships, boats, offshore structures and other deep-ocean installations. Areas covered in this second volume include design, the role of adhesives, fabrication techniques, and operational aspects such as response to slam loads and fatigue performance. The final three chapters of the book cover regulatory aspects of design, quality and safety assessment, and management and organisation.

These volumes will provide an up-to-date introduction to this important and fast-growing area for students and researchers in naval architecture and maritime engineering. It will also be of value to practising engineers as a comprehensive reference book.

CAMBRIDGE OCEAN TECHNOLOGY SERIES 5

COMPOSITE MATERIALS IN MARITIME STRUCTURES

Volume 2: Practical Considerations

Cambridge Ocean Technology Series

1. Faltinsen: Sea Loads on Ships and Offshore Structures
2. Burcher & Rydill: Concepts in Submarine Design
3. Breslin & Andersen: Hydrodynamics of Ship Propellers
4. Shenoi & Wellicome (eds.): Composite Materials in Maritime Structures, Volume 1, Fundamental Aspects
5. Shenoi & Wellicome (eds.): Composite Materials in Maritime Structures, Volume 2, Practical Considerations

COMPOSITE MATERIALS IN MARITIME STRUCTURES

Volume 2: Practical Considerations

Edited by

R. A. Shenoi

and

J. F. Wellicome

Department of Ship Science, University of Southampton

West
E uropean
G raduate
E ducation
M arine
T echnology

CAMBRIDGE UNIVERSITY PRESS
Cambridge, New York, Melbourne, Madrid, Cape Town, Singapore, São Paulo, Delhi

Cambridge University Press
The Edinburgh Building, Cambridge CB2 8RU, UK

Published in the United States of America by Cambridge University Press, New York

www.cambridge.org
Information on this title: www.cambridge.org/9780521451543

First published 1993
This digitally printed version 2008

A catalogue record for this publication is available from the British Library

ISBN 978-0-521-45154-3 hardback
ISBN 978-0-521-08994-4 paperback

CONTENTS

PREFACE

The use of Fibre Reinforced Plastics (FRP) in the marine field has been growing steadily since the early 1950s. Initially FRP was used for small craft such as lifeboats and pleasure craft. This has changed over the years to the point where structures having a mass of several hundred tonnes are regularly produced and used. Potential applications range from small components such as radar domes, masts and piping to larger scale structures such as ship hulls, ship superstructures, submersibles and offshore structure modules.

Alongside this growth in the number and size of FRP applications have come advances in materials technology, production methods and design procedures. It is now possible in many instances to produce structures which out-perform metal structures in terms of weight, strength and cost. To achieve this performance, good quality control procedures are vital. Equally, it is necessary for the designer and producer to have an all-round knowledge of FRP composite materials and related mechanics.

This book and its companion volume, which deals with Fundamental Aspects, are intended to provide a sound, theoretical base for the design and manufacture of major load-bearing structural members fabricated from FRP composites and to illustrate, through case studies, the particular features of the use of FRP in the marine field. The material has been grouped together to form two companion volumes that may also be useful independently of each other.

Volume 1 is titled "Composite Materials in Maritime Structures - Fundamental Aspects". This book contains the fundamental materials sciences, a discussion of failure mechanisms in FRP and the theoretical treatment of failure in design. There is a discussion of the applied mechanics of complex, orthotropic laminates and the basis of strength calculations for single skin and sandwich structures. Attention is also given to numerical techniques such as finite element analysis.

Volume 2 is titled "Composite Materials in Maritime Structures - Practical Considerations". This book is devoted to applications of FRP in the maritime field. It has sections on the design of craft operating in a displacement mode or with dynamic lift and support. Production considerations related to single skin and sandwich structures are discussed. Operational aspects related to material and structural failure are covered. The book closes with an examination of the impact of regulations on quality and of the complexities of design management.

The authors of various chapters in the two books are all professionally engaged in the field of FRP composites technology working in the marine industry, classification societies, research establishments and universities. They are known internationally for their contributions to the subject. Without the tremendous effort put in by the authors and their cooperation in meeting deadlines, these books would not have been possible. The editors wish to thank the authors for all their help and assistance.

The material in these two books was compiled for the 18th WEGEMT Graduate School held in the University of Southampton in March 1993. WEGEMT is a European network of universities in Marine Technology which exists to promote continuing education courses in this broad field, to encourage staff and student exchanges among the 28 (current) members and to foster common research interests.

Participants at such schools have generally been drawn from the ranks of practising, professional engineers in shipyards, boatyards, offshore industries, design consultancies and shipping companies. A large proportion have also been postgraduate students and staff from various academic and research establishments wishing to obtain an overview of a particular topic as a basis for research.

It is intended that these two books should be of interest to a similar spectrum of readers and that they could be used as a text for advanced undergraduate or postgraduate courses in FRP technology. Furthermore, they could be used as a basis for continuing education courses for young engineers in industry. Although aimed primarily at the marine field, the material in these books would also be relevant in other contexts where FRP is used for structural purposes.

The 18th WEGEMT School was organised with the help of an International Steering Committee whose members were:

Mr. O. Gullberg	Karlskronavaarvet AB, Sweden
Mr. A. Marchant	AMTEC, UK
Prof. M.K. Nygard	Veritas Research & University of Oslo, Norway
Prof. H. Petershagen	University of Hamburg, Germany

Dr. R. Porcari	CETENA, Italy
Ir. H. Scholte	Delft University of Technology, The Netherlands

The Committee approved the course content and helped select the course lecturers whose notes form the material of these two books. The editors are grateful for their advice and assistance. The School was supported in part by COMETT funds which were obtained from the EC by the University Enterprise Training Partnership (Marine Science & Technology) which is administered through the Marine Technology Directorate Ltd., London. We are indebted to Mr. J. Grant, Secretary General of WEGEMT for his help in obtaining COMETT funding and in publicising the School.

The encouragement, support and assistance given to this venture by Professor W.G. Price and our other colleagues in the Department of Ship Science has been incalculable and valuable. We are grateful to them. Finally, we extend our thanks to Mrs. L. Cutler for her expertise and professionalism in undertaking the word processing and for patiently coping with the numerous edits, changes and amendments involved in preparing the camera-ready copy for the two books.

R.A. Shenoi, J.F. Wellicome
Southampton

LIST OF AUTHORS

- Professor H.G. Allen, University of Southampton, U.K. (Vol. 1)
- Dr. J. Benoit, Bureau Veritas, Paris, France. (Vol. 2)
- Mr. A. Bunney, Vosper Thornycroft, Southampton, U.K. (Vol. 2)
- Dr. D.W. Chalmers, Defence Research Agency, Haslar, U.K. (Vol. 1)
- Dr. P.T. Curtis, Defence Research Agency, Farnborough, U.K. (Vol. 1)
- Dr. R. Damonte, CETENA, Genova, Italy. (Vol. 1)
- Mr. A.R. Dodkins, Vosper Thornycroft, Southampton, U.K. (Vol. 2)
- Professor A.G. Gibson, University of Newcastle-upon-Tyne, U.K. (Vol. 2)
- Professor K. van Harten, Delft University of Technology, The Netherlands. (Vol. 1)
- Mr. G.L. Hawkins, University of Southampton, U.K. (Vol. 2)
- Dr. B. Hayman, Det Norske Veritas, Hovik, Norway. (Vol. 2)
- Mr. S-E. Hellbratt, Kockums AB, Karlskrona, Sweden. (Vol. 2)
- Mr. P. Krass, Schutz Werke, Selters, Germany. (Vol. 2)
- Mr. A. Marchant, AMTEC, Romsey, U.K. (Vol. 2)
- Professor G. Niederstadt, DLR, Braunschweig, Germany. (Vol. 1)
- Ir. A.H.J. Nijhof, Delft University of Technology, The Netherlands. (Vol. 1)
- Dr. G. Puccini, CETENA, Genova, Italy. (Vol. 1)
- Mr. R.J. Rymill, Lloyd's Register of Shipping, London, U.K. (Vol. 2)
- Ir. H.G. Scholte, Delft University of Technology, The Netherlands. (Vol. 2)
- Dr. R.A. Shenoi, University of Southampton, U.K. (Vols. 1 & 2)
- Dr. G.D. Sims, National Physical Laboratory, Teddington, U.K. (Vol. 1)
- Mr. P.J. Usher, Vosper Thornycroft, Southampton, U.K. (Vol. 1)
- Mr. I.E. Winkle, University of Glasgow, U.K. (Vol. 2)

1 INTRODUCTION

This book deals mainly with practical issues related to the application of FRP composite materials in a maritime engineering context. The contents may be seen to fall under four, broad categories namely design aspects, production considerations, in-service or operations-related features and safety/quality management issues.

The next two chapters, related to the design of displacement vessels and dynamically-supported craft, emphasise the centrality of design. The two vessel types - a minehunter and a Surface Effect Ship - cover the extremes in terms of loading, structural topology, material configurations and operational scenarios. Whereas these chapters cover the overall aspects of preliminary design, the next two (Chapters 4 and 5) on adhesives and joints serve to underline the importance of detail considerations that need to be given to secondary bonding in FRP structures.

The second category of topics covered in the book relates to production factors. Again, an attempt has been made to cover the two "extremes" related to single skin stiffened and monocoque sandwich structures (Chapters 6 and 7). The former is the traditionally derived form for high quality naval vessels while the latter is increasingly favoured for smaller, more specialist craft such as yachts for America's Cup races.

The third category is related to operational considerations. Owners and operators of craft need to know the significance of defects such as cracks and blisters, as and when they appear, on performance and safety (Chapter 8). There is also a need to have an adequate understanding of the response of FRP structures to typical marine loadings and of possible failure patterns under large loads (Chapter 9). Lastly, in this category, because the life of a typical commercial vessel can be up to 25 years, it is important to understand long term performance characteristics of the material and structural configurations (Chapter 10).

The final category of topics is related to more general but vital issues affecting design of maritime FRP structures. Whereas the majority of structural applications in the maritime field up to now have been restricted to ships, boats and submersibles, there is a growing use of FRP for structural applications in the offshore field. There is a need to highlight specific requirements in this context, such as fire risks, and to identify principal characteristics of design (Chapter 11). The International Maritime Organisation and regulatory authorities in different countries attempt to rationalise and standardise quality and safety aspects related to ships and other marine structures. It is essential that designers, producers and users of marine FRP structures are aware of the various statutes and regulations (Chapters 12 and 13). Finally, because of the complexities of design with FRP, where a vast amount of information is both required and generated at various levels of design, it is important to organise the task of design carefully (Chapter 14).

2 DESIGN OF DISPLACEMENT CRAFT

2.1 INTRODUCTION

2.1.1 Background

This Chapter considers the use of composite materials for the construction of displacement vessels such as mine countermeasures vessels (MCMVs). These are currently the largest ships afloat built entirely of fibre reinforced plastics (FRP). In the UK alone there are 19 FRP construction MCMVs of three successive classes, either at sea or under construction. The largest FRP ships are the USA Osprey Class MHC, having an overall length of 63 metres and a deep displacement of 850 tonnes.

A variety of structural `styles' are used, including stiffened single skin, sandwich, frameless monocoque or a combination of all three. Corrugated construction has been suggested as a possible structural style for future generations of MCMVs.

2.1.2 Why Composite Materials?

GFRP composite materials have been the clear choice for the construction of MCMVs of 45-60 metres in length for the last 25 years, largely because of characteristics other than their mechanical properties, and have in fact replaced wood almost completely. For MCMVs composites offer the following advantages over steel and wood, enabling the exacting design requirements of these specialised ships to be met:

- Non-conducting
- Non-magnetic
- More economical to build and repair than wood
- Better ruggedness and explosion resistance than wood
- Does not rot or corrode in a marine environment.

In considering the use of FRP for the construction of other, perhaps more conventional types of ships, mild steel remains the most obvious choice. One of the main advantages offered by composite materials over steel is weight-saving. However, although weight-saving in any type of marine structure is an advantage, it is least critical in the case of displacement vessels. Perhaps the first application on a ship without the special material requirements of MCMVs will be for superstructures and masts, so that weight-saving can also be turned into reduced top weight and improved stability.

The corrosion resistance of FRP is potentially one of its greatest advantages and one which the chemical industry has not failed to recognise. FRP ought to be a prime candidate for vessels intended to operate in particularly severe environments such as the Middle East, where corrosion rates are rapid and the cost of continuous maintenance and repair of steel hulls is prohibitive.

Considering construction cost, FRP and steel are roughly equal for vessels of around 25 metres in length if tooling is excluded from the calculation. Smaller vessels tend to be more economical to produce in composites and larger vessels in steel.

2.1.3 Subject of Case Study

After discussing some of the important fundamental considerations of structural design philosophy, the design of a modern 50 metre FRP displacement vessel of stiffened single skin construction will be addressed, and the UK Sandown Class minehunter will be used as an example to illustrate many of the points being made.

In the remainder of this Chapter, the following points are addressed:

i) The selection of materials and methods of material characterisation.
ii) The selection of structural style.
iii) The influence of features of the ship's general arrangement on the structural design.
iv) The influence of production constraints on the structural configuration.
v) The design and analysis of the ship's structure.

2.2 MATERIAL SELECTION AND TESTING

2.2.1 General Requirements

The vessel size places a significant constraint on the materials which can be used. To minimise the cost of tooling the method of construction is limited to the open-mould wet lay-up process. Curing must take place in air at ambient temperatures (15-25°C). From a practical point of view this precludes all resins except polyester and possibly vinyl ester. Of these two polyester is the cheaper, easier to use, gives acceptable properties and has excellent resistance to the marine environment. For the reinforcements, the choice is between E-glass, R-glass, carbon and aramid. E-glass is several orders of cost cheaper than any of the others, results in an acceptable structural weight (at least for displacement vessels), is easy to tailor, has good fire resistance and good all-round mechanical properties. E-glass reinforcements in a polyester resin matrix is therefore the logical choice, whatever style of construction is finally chosen, although other materials may be specified for special applications where the need to save weight is more important than the increased cost, ie. for mast construction or for ballistic resistance. An interesting trend is the use of hybrid reinforcements which combine E-glass with aramid fibres. Here a weight-saving of 15% can be made without significantly increasing cost.

Reinforcements can be in the form of random fibre mat, woven rovings, stitched multi-axial or knitted fabrics. For shipbuilding woven rovings and combination cloths (stitched woven and mat layers) have tended to dominate, but increasing interest is being shown in multi-axials, which eliminate the problem of fibre crimp which is characteristic of woven fabrics.

The properties of laminate density, thickness, strength and moduli depend largely on the properties of the constituent reinforcement and matrix materials from which they are created, but they are also modified by the conditions under which laminating takes place. The type of laminating process, the overall size and nature of the structure, the level of workmanship and supervision and even the ambient temperature all conspire to make theoretical property prediction difficult. Practical laminating trials in the proposed building environment and characterisation testing are therefore vital to obtain reliable values for design purposes.

2.2.2 Production Considerations

The characteristics of laminating materials affect both the quality of the final laminate and the production time. These features may be evaluated realistically only by large scale production trials. Important aspects to consider are:

(a) Ease of cutting of E-glass cloth is largely dependent on the weight of the cloth. If it is intended to use a range of standard widths, e.g. for stiffener over-laminating it is an advantage to have the material supplied ready cut to those widths with the edges stitched to prevent fraying. Woven roving is more likely to fray than combination cloth where the stitching and mat layer hold the cut edges.

(b) A low viscosity resin reduces the time taken for consolidation (wet-out of the cloth and removal air bubbles by rolling).

(c) The laminating resin should be thixotropic to reduce drainage on vertical surfaces. However this tends to increase the difficulty of wetting-out. It should be noted that while a tightly packed woven roving cloth may be laid-up with a high resin content in a test on a horizontal panel, it may drain badly in a vertical lay-up test.

(d) While most cloths will wrap easily around a cylindrically shaped mould surface, not all will form around a corner or a shape with double curvature without some tailoring. Examples are the snaped ends of stiffeners, tapered stiffener sections and the bow section of a hull. Cloths with poor drapability should be avoided as they lead to excessive tailoring which results in cloth joints which are too close together, necessitating additional material to compensate for the loss in strength.

(e) Conversely a cloth which is too drapable is too easily distorted such that the rovings are pulled out of a straight line. Coloured threads may be incorporated with the warp rovings to help maintain straightness during lay-up. Combination cloths are kept straight by the mat layer and stitching.

(f) Time is not necessarily saved by using a heavy cloth to reduce the number of plies laid. Each cloth will take longer to consolidate and in any case there is a limit on the weight of cloth and resin that can be laid wet-on-wet at one time (see below).

(g) The curing reaction of polyester resin is exothermic. If the laminate thickness build-up is too fast the later layers tend to insulate the earlier layers and prevent the dissipation of heat. The laminate then begins to heat up more which further accelerates the curing reaction until a runaway situation develops and the temperature may rise to the point where the laminate becomes permanently heat-damaged.

(h) Practical tests should be carried out to assess the ease at which step cutting for scarf joints may be carried out.

2.2.3 Production Trials

Initial trials may be carried out on small panels measuring about 1 m x 2 m. These should have a stiffener bonded to the surface to check drapability over the sides and ends of the section. Handling may be evaluated to a limited extent and resin ratio and thickness per ply checked. At this early stage in the material selection process a large number of materials may be evaluated economically.

Having short-listed the materials with adequate handling characteristics, panels of about 3 m x 3 m x 8 mm are required, from which samples can be cut to determine mechanical properties. Panels of this size are important as they reflect the level of difficulty involved in large scale production laminates and will include a realistic void content and butted and staggered cloth edges.

Once mechanical tests have been completed the number of candidate cloth/resin combinations may be reduced to perhaps two or three. These materials may now be tested again in more realistic production trials. Short sections of hull, perhaps three or four frame spaces long as a minimum, should be laid-up in a hull mould. These sections should include examples of all principal structural features of the proposed design, such as frames, longitudinals, bulkheads, tee-joints, floors, decks, beam knee joints and other stiffener intersections. The various points discussed under Production Considerations may be examined. Samples should be cut from various parts of the section for confirmatory tests of resin ratio, thickness, and mechanical properties.

2.2.4 Mechanical Tests and Calculation of Properties

Mechanical tests may be carried out using naval standards or BS, ISO, European or ASTM. Some recommended standards are:

Tensile	BS 2782 Part 10 Method 1003
Flexural	BS 2782 Part 10 Method 1005
Short beam shear	BS 2782 Part 3 Method 341A
Compression	BS 2782 Part 3 Method 345A
In-plane shear	ASTM D4255 (1983) Method A
Poisson's ratio	Incorporated with Tensile test
Degree of cure (Barcol Hardness)	BS 2782 Part 10 Method 1001
Preparation of specimens	BS 2782 Part 9 Method 930A by machining

Laminates incorporating woven reinforcement are tested in both warp and weft directions. A minimum of five specimens should be prepared for each test to obtain a realistic picture of the likely scatter in properties. Thus for tensile strength and modulus in the warp direction, five warp oriented specimens should be cut from each of vertical and horizontal panels. For balanced weave fabrics the warp and weft results may be combined. The results of mechanical tests should be analysed by conventional statistical methods.

The values of strength and modulus to be used in the structural design calculations should be obtained by taking the mean values and deducting twice the standard deviation. This gives a 97.5% probability that the design figure will be exceeded in the actual structure. It must be remembered that large scatters are an inherent feature of composite testing.

2.2.5 Long-Term Properties

All resins absorb a small percentage (around 1% in the case of polyester) of moisture. The general pattern of degradation in a marine structure is an initial fall of 10-20% in mechanical properties in the first year to 18 months followed by a very slow fall over the rest of its life. However, if a saturated laminate is subject to continuous tensile stress the degradation may be more severe. Laboratory experiments have shown that the sustained stress levels should not exceed 10-20% of the ultimate values. This is generally not a problem in ships' structures designed for extreme load cases due to wave action etc, but special attention should be paid to structure supporting dead loads. It is advisable also to keep strain levels below those at which resin microcracking occurs, say 0.3%-0.5%, thus avoiding increased moisture ingress.

Accelerated ageing tests should be carried out on laminate specimens which are either immersed in water or placed in a condensation chamber. The ageing process is accelerated by either increasing temperature or pressure above ambient. For correlation of results to real time, the test temperature should be kept below the heat distortion temperature. This is typically 60-70°C for a non-post cured polyester. Quicker results may be obtained by heating to 80-90°C and comparing the performance to that of a laminate of known resistance to ageing. The standard test used by the UK MOD consists of immersing the specimens in fresh water at 23°C and 7MPa for a period of 100 days.

In specifying the shell lay-up of a hull an all woven roving lay-up is acceptable, but the outer surface must contain a layer of chopped strand mat. The mat takes up a greater proportion of resin than woven rovings and the short-fibres ensure than if the glass becomes exposed on the surfaces, water cannot wick far along the fibres. The mat layer therefore forms a protective barrier to the underlying woven layers. The use of gel coats on yachts perform the same function (as well as giving a good aesthetic finish).

2.3 DESIGN CRITERIA

2.3.1 General

The design of a ship's primary hull girder structure to take account of bending and shear effects due to static and dynamic wave loading and also, in the case of MCMVs, explosion induced whipping loading, involves performing the following analyses:

(a) Calculation of local panel and stiffener scantlings and hull section moduli such that longitudinal bending moments and shear forces in combination with lateral hydrostatic pressures do not cause material stresses in excess of their permissible values. In this analysis allowance must be made for stress concentrations created by structural discontinuities in the hull girder (particularly main deck) structure.

(b) A check to ensure that adequate hull girder natural frequencies are obtained and that deflections under a variety of loading conditions are not excessive.

(c) A calculation of the buckling resistance of local unsupported panels, individual stiffener elements and combined stiffened panels between main bulkhead supports. Such calculations are particularly important for hull bottom and main deck structure under compressive loading, due to the relatively low Young's Modulus of FRP materials normally used for construction.

The above philosophy therefore differs from that applied in small boat structural design, where the choice of shell scantlings is usually determined by hydrostatic and hydrodynamic pressures, as well as a general ruggedness requirement.

2.3.2 Design Margins

For most marine applications where woven or mat reinforcements are used, it is reasonable to use isotropic and orthotropic plate theory for design, although it must be remembered that the materials are essentially anisotropic in nature and a thorough understanding of modes of failure is essential.

The following design margins are typical of current practice. The values shown make no allowances for loading or material uncertainties, ie. maximum loads and minimum mechanical properties:

Static short-term loads (tension)	3.0
Static short-term loads (compression)	2.0
Static long-term loads (dry)	4.0
Static long-term loads (immersed)	6.0
Load reversal	5.0
Local buckling (stiffeners parallel to load)	1.5
Column buckling of stiffener/plate combinations	2.0
Buckling (stiffeners perpendicular to load)	3.5

The margins for short-term static loads are those which, if applied to the ultimate strength of the laminate, will give the resin microcracking stress. Stresses higher than the microcracking stress will cause permanent damage to the laminate, although it will carry on taking load up to the ultimate value.

In the case of local buckling of panels between stiffeners, the low margin of 1.5 is only justified if positive measures are taken to prevent premature detachment of stiffeners from panel. This may involve bolting of flanges to the panel or using a resilient adhesive which inhibits peeling of the flanges.

Regarding fatigue, the margin of 5.0 applied to ultimate stress only relates to high strain rate load application. Static short-term margins may be used for the overall hull structure without risk of fatigue failure.

Bearing in mind the low modulus of FRP it is important to evaluate structural deformations. When designing tanks to withstand internal pressure a deflection limit of L/200 should be imposed to avoid excessive deformation and possible subsequent damage to boundary joints. Panels between stiffeners on lightweight decks, for example in the superstructure, should be limited in deflection to b/80 (where b is the panel width) to avoid them feeling springy to walk on, for although strength may be adequate this phenomenon can be somewhat disconcerting to the ship's staff!

2.3.3 Loads

As loads are independent of the structural materials, loading on the structure will not be addressed in this Chapter. However, the relative importance of in-plane loading from the point of view of buckling performance is emphasised. A description of load cases is given in table 2.1.

Table 2.1. Loads imposed on ship structures.

Basic Loads	Sea Loads	Operational Loads	Combat Loads
Live Loads	Hull Girder Bending	Flooding	Primary Shockwave
Structure Self Weight	Wave Slamming	Helicopter Landing	Explosion induced Whipping
Tank Pressures	Hydrostatic/ Wave Slap	Docking	Gun Blast Pressures
Equipment Weights	Roll/Pitch/ Heave Inertia	Replenishment at Sea	Fragmentation
	Wind	Anchoring	Gun Recoil
		Berthing	Missile Efflux Pressures and Hang-Fire Case

2.4 STRUCTURAL SYNTHESIS

2.4.1 Structural Style

There are some radically different structural styles from which to choose, see figure 2.1, each having its own advantages and disadvantages when considering their application to large displacement vessels. These are summarised in table 2.2 and relative weights and costs of these different forms of construction are given in table 2.3.

Table 2.2. Comparison of structural styles [4].

Advantages	Disadvantages
(a) Top Hat Stiffened Single Skin Structure	
Properties and responses are well-known Automation possible Easy to fit equipment Costs reduce with number of hulls Quality control is easy Survey in service is straightforward	Fairly expensive to build Care is needed to provide good impact resistance
(b) Monocoque Structure	
Easily automated Low labour cost Few secondary bonds below waterline Good shock resistance	Very heavyweight High material cost Survey methods need development Attachments and support to machinery are difficult Quality control difficult
(c) Sandwich Structure	
High bending stiffness for low weight can be achieved Can be built without a mould Secondary bonding can be minimised Construction, maintenance and operating costs are lower than conventional structure Easy to fit equipment	Survey methods need development Long-term durability is not proved Special precautions need to be taken to protect the core from fire damage
(d) Corrugated Structure	
Relatively lightweight Low material and labour cost Automation is possible	Lower transverse strength Fitting internal structure may be difficult Awkward mould Strange appearance

It may be concluded from table 2.2 that

(a) Sandwich construction offers a fairly low cost (at least for one-offs or small production runs) and high stiffness-to-weight ratio, at the expense of service durability, ease of Quality Assurance (QA) inspection and maintenance, although these aspects are being successfully addressed in small craft construction and will no doubt improve for larger vessels in time as more experience is gained.

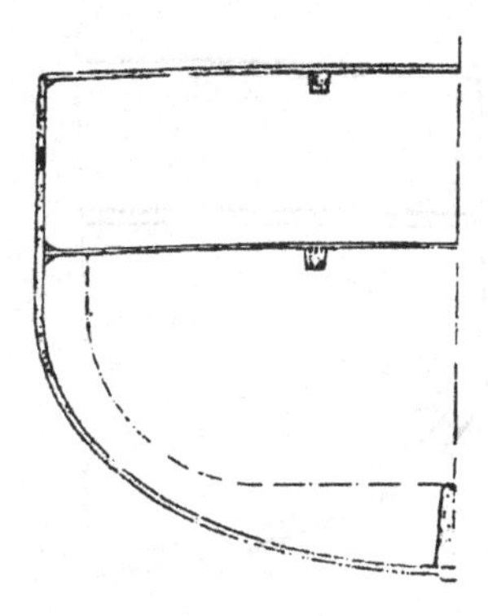

Figure 2.1(a). Sandwich construction hull.

(b) With a large capital investment monocoque construction may be mechanised to a very large extent, thereby minimising labour cost, but results in a very heavy structure. It is best suited to very long production runs and where vessel weight is no object. QA during construction and inspection in-service can also be difficult.

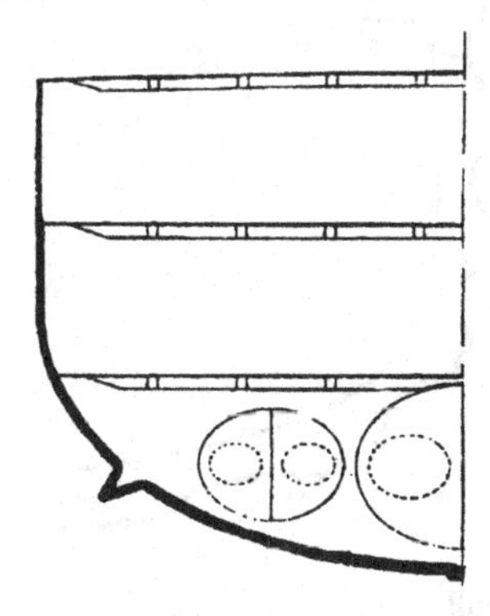

Figure 2.1(b). Monocoque hull.

(c) Stiffened single skin construction offers the lowest technical risk in that design, build inspection, maintenance and repair are all straightforward. Therefore where strength and durability under a variety of loading conditions has to be very good and weight has a slightly lesser emphasis, single skin construction is particularly suitable for displacement vessels. Cost is higher than sandwich, especially for one-offs, but this difference is reduced when production runs of five or more vessels are planned.

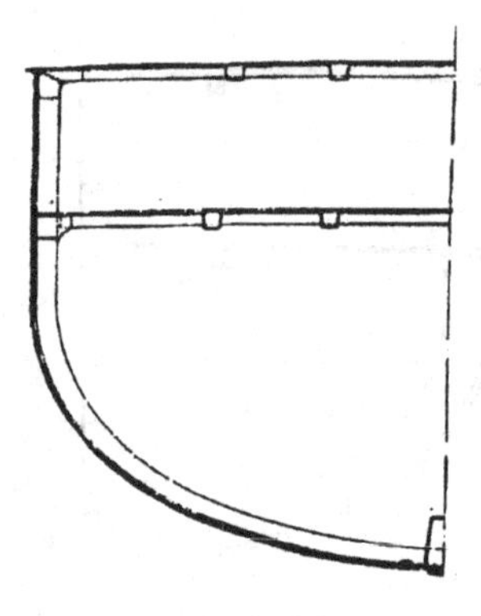

Figure 2.1(c). Transversely framed single skin hull.

(d) Corrugated construction offers lighter weight than stiffened single skin, but is unlikely to be seriously considered for hull structures without considerable further development. It is relatively expensive, particularly in terms of tooling cost and lay-up complexity.

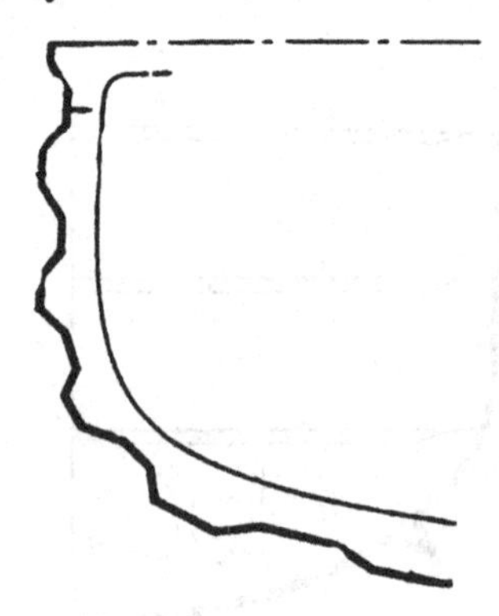

Figure 2.1(d). Corrugated hull.

It is possible to mix these different forms of construction to combine the advantages and obtain the best compromise for a particular application. For example, it may be attractive to specify a single skin hull and main deck, corrugated watertight bulkheads and sandwich construction for secondary structure such as internal decks, minor bulkheads and superstructure. The UK Sandown Class minehunter and the French BAMO minehunter use two or more construction styles.

Table 2.3. Comparison of weights and costs [8].

Configuration	Relative Weight	Relative Cost
Single Skin		
Longitudinally stiffened bottom with framed sides	1.00	1.00 (0.75)*
Transversely framed bottom and sides:- Frames at:		
- 1 m spacing	1.23	1.17
- 1.67 m spacing	1.53	1.24
- 2.5 m spacing	1.82	1.26
Orthogonally stiffened bottom and sides frames at 1 m spacing Longitudinals at:		
- 1 m spacing	1.10	1.63
- 2 m spacing	1.08	1.60
Faced Corrugations		
0.3 m depth	0.75	0.79
Corrugations		
0.16 m depth	1.24	1.55
0.48 m depth	0.75	0.94
PVC Foam Core Sandwich	0.73	0.62
Solid Thick GRP (unstiffened)	3.04	1.92

* Compliant resin used instead of bolts

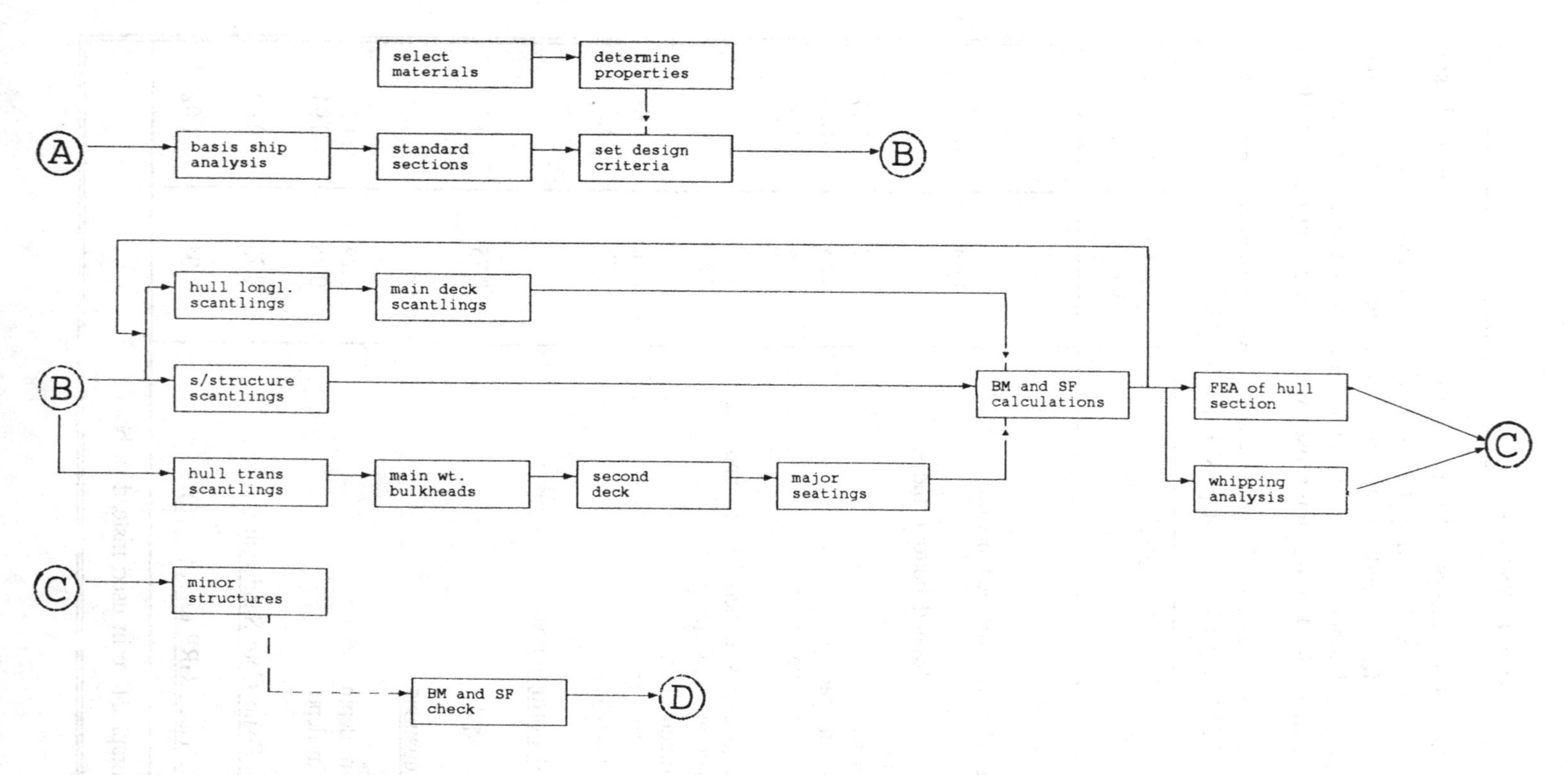

Figure 2.2. Design process for GRP MCMVs.

2.4.2 Synthesis Modelling

2.4.2.1 Overview of design process

The overall design process applicable to large displacement vessels is illustrated in figure 2.2. This is similar to steel ship design, with a number of additional analysis tasks arising from the type of material and the requirement for shock resistance in the case of MCMVs.

Throughout the design process attention must be paid to structural continuity, stiffener and tee-joint attachments and terminations, as well as the provision of support for machinery and equipment.

2.4.2.2 General arrangement - influence on structure

Certain features of the ship's general arrangement can have a marked influence on the complexity, and hence the cost, of the structure. In particular the following points should be noted during the early design stages in order to keep the structural arrangement as simple as possible.

(a) Major transverse bulkheads should be placed in positions of multiples of the frame spacing. This avoids the complication of varying frame spacing along the length of the ship or landing bulkheads on the shell in positions too close to the existing frames.

(b) Bulkhead spacing should lead to approximately equal compartment lengths along the length of the ship. This is not always practical to achieve, but in extreme cases of long compartments adjacent to short ones, it may be necessary to taper longitudinal stiffeners, resulting in a high labour effort to shape foam formers and tailor lay-up cloths. It is preferable to maintain a constant former section and accommodate reasonable variations in spans by varying the lay-up alone. One exception is likely to be the machinery space, where longitudinals have large spans, but in any case need to be specially shaped to provide engine and gearbox foundations.

(c) It is not essential to position main transverse bulkheads at either end of the lower tier of superstructure. The flexibility of the FRP material will ensure that there are no significant stress concentrations at these points.

(d) In optimising transverse and longitudinal frame spacing, it is important to consider the space between stiffeners required for bolted skin fittings as well as ensuring good access to all stiffener surfaces for laminators. This means a frame spacing of around 1-1.5 metres for hull and main deck and 0.6-1.0 metres for superstructure and internal structure. If too high a stiffener spacing is selected the section sizes will be such that they begin to encroach too much into the useable compartment space.

(e) All minor bulkheads should be made structural so that they support the deck above. It is an advantage to keep passageways straight, as bulkheads then provide continuous and rigid lines of support for deck longitudinals and beams.

(f) The main deck is required to have a number of hatches and shipping openings. These should be confined to the centre of the ship and kept in line as far as possible. Thus longitudinals can run straight and parallel to the centreline outside the line of openings, with transverse beams running between inner longitudinals to provide local support to the edges of openings. This maximises the longitudinal section modulus and avoids `cranking' of longitudinals around openings, which adds to build complexity and reduces laminator productivity.

(g) In positioning deck or bulkhead penetrations allowance should be made for tee-joints at bulkhead to shell and bulkhead to deck connections. Penetrations should be kept clear of these joints, although bonding angles can be through-bolted to provide a strong attachment point and also serve to clamp the bonding angle to the plating.

(h) A unique property of composite materials is that almost any desired shape of structure may be produced by the use of an appropriately shaped mould. However, for simple structures such as ships superstructures, it can be considerably less costly to build them from an assembly of flat panels than to build them as a one-piece moulding. Some compromises, after carefully conducted production engineering studies, may therefore be necessary. The design of major structural components should maximise the use of flat panels, which can be built to any size on a general purpose flat panel tool. Any part having a 3-D shape requires a special mould to be built. For example, decks should ideally have no camber and only straight line sheer. This enables all panels to be made flat, avoiding the need for adjustable panel tools and specially shaped stiffener bases. It also enables all surfaces to be kept level, facilitating block construction and advanced outfitting.

(i) Integral tanks make maximum use of internal space. Major bulkheads and lower deck form tank ends and tank tops respectively. Longitudinal tank bulkheads should be so arranged as to land on the centre of longitudinal bottom stiffeners.

2.4.2.3 Structural arrangement

Figure 2.3 shows the structural arrangement of the Sandown Class minehunter. This differs from that of the previous Hunt Class MCMV in that it features a longitudinally stiffened bottom and main deck structure. This is a result of feasibility studies for the Sandown Class which showed that a longitudinally stiffened structure would be more efficient than a transversely stiffened one in terms of both weight and construction cost for the following reasons:

(a) More of the structure is effective in resisting longitudinal hull girder bending.
(b) Stiffener intersections are greatly reduced.
(c) Factors of safety against buckling of the shell plating may be reduced, which is justified by the reserve of strength present in the stiffening.

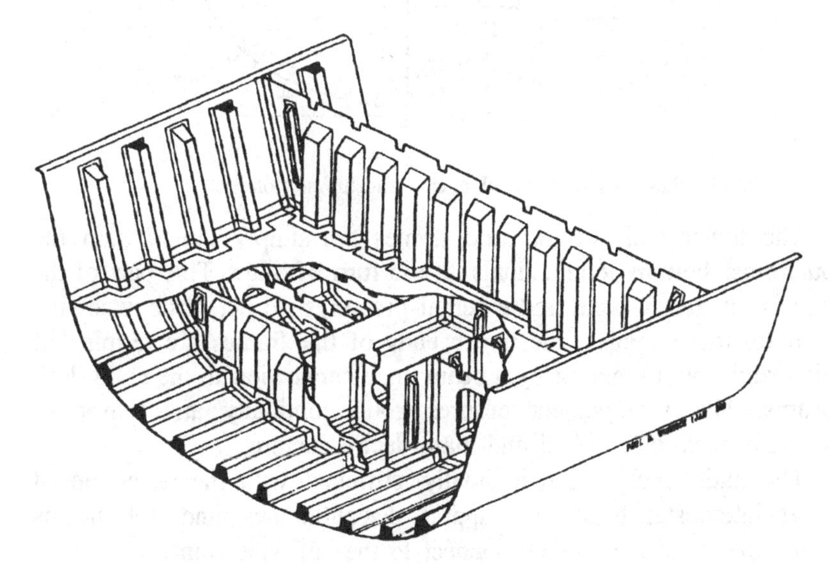

Figure 2.3(a). Longitudinally stiffened single skin hull (Sandown class minehunter).

The disadvantages of longitudinal framing are that stiffener bases must be shaped to land upright on the varying deadrise angle of the ship's bottom; the difficulty of laminating longitudinals on the side shell (hence side frames are vertical); and an increase in weight of main transverse bulkheads to provide support for the longitudinal stiffeners (hence these are corrugated to minimise weight).

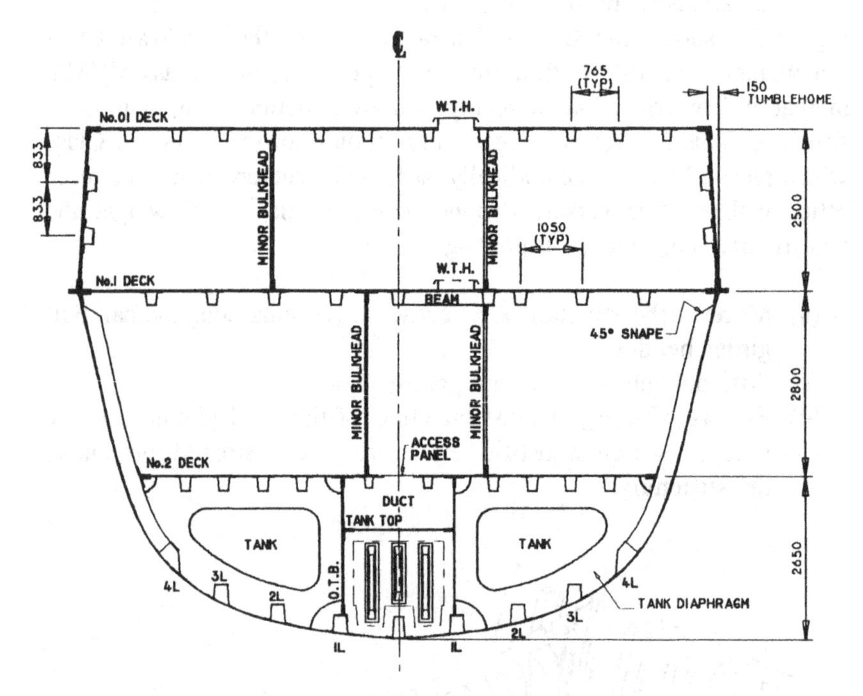

Figure 2.3(b). Sandown class minehunter - midship section.

The lower ends of the side frames are simply butted onto the outermost bottom longitudinals at the turn of bilge. This part of the hull is inherently rigid and external pressures do not place excessive load on these joints. The upper ends of the frames are terminated alternately by a snape or by a beam knee connection to the main deck beams. This arrangement ensures good continuity and transverse strength between the hull and main deck structure.

The main deck is longitudinally stiffened with the exception of short intercostal beams to support discontinuities and stub beams along the deck edge which connect to the hull side frames.

Shell plating thickness is on average just over 20 mm, with extra shell reinforcement placed locally in way of tanks, slamming region and other highly loaded points. Deck plating thickness ranges from

15-25 mm, reflecting the variation of longitudinal loading and lateral pressure along the length of the hull.

2.4.2.4 Plating design

The basic bottom shell thickness is determined by the requirement for it to resist a combination of in-plane loading and lateral pressures. Initially bottom plating should be considered as long flat isotropic plates, fully fixed along stiffener centrelines for the general hydrostatic pressure case and orthotropic and pinned along the stiffener web to shell intersection for the case of local buckling between stiffeners.

Where lateral pressure is found to dominate plating thickness a simple finite element model of a representative stiffener and panel combination and ply-by-ply laminate failure analysis may be undertaken to further refine the scantlings. However, this technique cannot be used until initial scantlings have been determined by simple beam and plate theory.

Having established side shell plating thickness under lateral pressure, an orthotropic plate finite element model should be developed of a typical three bay frame and shell panel to examine resistance to buckling under combined in-plane shear and linearly varying longitudinal in-plane compression loading resulting from the hull girder sagging case.

2.4.2.5 Stiffener design

For stiffeners the designer has to select from the following variables:

- Section height
- Section width
- Web angle
- Flange width
- Web lay-up
- Table lay-up

The almost infinite freedom that this represents is tempered by the need to standardise as much as possible. The penalty for not doing so is to give the production department an almost impossible task in shaping foam former sections, tapering from one size to another and tailoring cloth widths to suit the varying section sizes. Too many changes in number of lay-up plies along the run of one stiffener can also cause many problems to the laminator. Overall the result can be a dramatic reduction in productivity.

The best approach is to devise a range of standard section sizes, preferably 10 different sizes or less for a ship, and try to live with one particular size of section for the length of run of each stiffener. If weight-saving is important then it is quite acceptable to vary the lay-up along the length of the stiffener. For example a longitudinal hull bottom stiffener may run through several compartments, being supported by main transverse bulkheads in positions such that it is divided into several spans of unequal length. By selecting the right section from the standard range, it should be possible to cope with the different spans by varying the lay-up from one compartment to the next and achieve this without incurring a significant weight penalty.

Although full width reinforcements are usually laid over the foam former, it is often beneficial to either use two cloths per ply, overlapping them on the table, or preferably to incorporate strips of unidirectional reinforcement in the stiffener tables, to boost second moment of area and minimise the weight of the section.

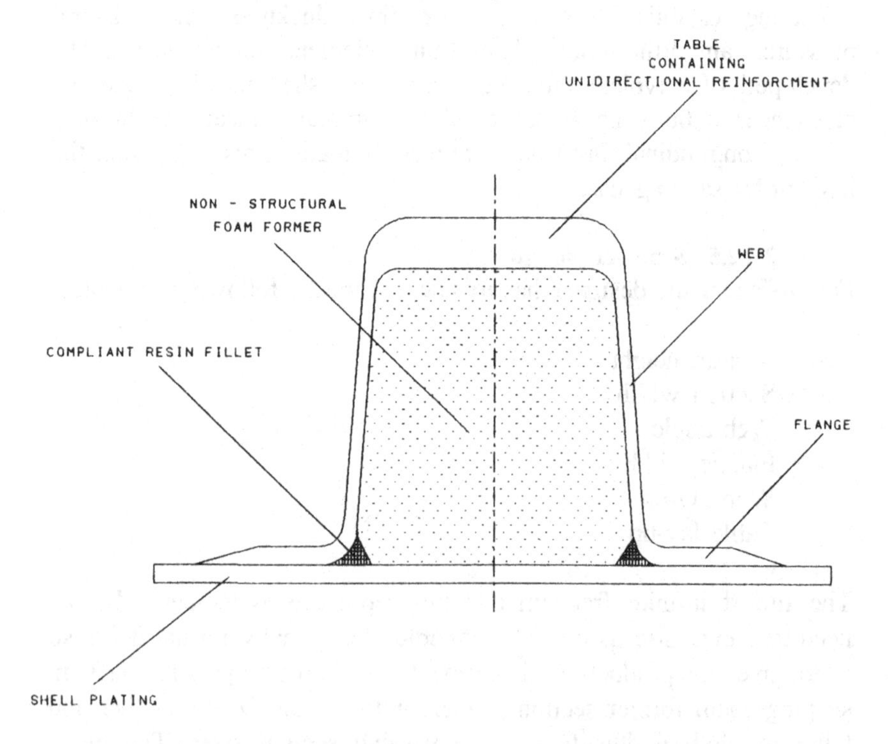

Figure 2.4. Single skin hull - typical stiffener to shell connection.

In design of hat-section stiffeners, see figure 2.4, where stiffened panels are subject to lateral and end loading the following modes of failure should be considered:

- Shear stress failure of webs
- Tensile/compressive stress failure of table
- Exceedence of strain limit in tables containing unidirectional reinforcement
- Tensile/compressive stress failure of base panel
- Exceedence of pre-determined central deflection limit
- Local compression buckling of table
- Shear buckling of webs
- Column buckling of stiffener/plate combination
- Interlaminar shear failure of flange to base plate bond

Once a first pass at stressing a stiffener has been carried out the most critical of the above modes of failure will become evident.

In calculations of section constants for a beam made of composite materials it is important to remember to account for differences in Young's Modulus and ultimate strength between base panel, stiffener and any unidirectional reinforcement.

2.4.2.6 Mathematical Modelling

To model a ship's hull, or even a section, using laminate finite elements, would be an extremely laborious task and would use a great deal of computation time. The preferred approach is to undertake FE analysis at two levels.

The global response of the hull structure can be modelled with sufficient accuracy using general purpose codes and isotropic elements. This will give a realistic assessment of strain distribution around the hull section and deformation of hull and deck panels between supporting bulkheads.

To examine stress distribution at a detailed level, local models of stiffened panels can be created and boundary conditions determined from the global model can be applied. Orthotropic elements are adequate for this purpose.

From the local model the loading at any point within the stiffened panel can be obtained and a layer-by-layer laminate analysis carried out. This gives results in terms of stresses, strains and reserve factors for each layer as well as calculating the composite properties (modulus and stiffnesses) of the plate. The lay-up can be modified as desired and the results fed back into the global and local models to re-run the analysis.

There are special purpose codes and elements which are best used at a research level for modelling specific details of a structure such as joints. Results should always be validated by experimental data.

2.5 CONCLUSIONS

Developments over the last 25 years since the first GRP ship, HMS Wilton, was launched, have led to the availability of a wide choice of materials and fabrication techniques for ship construction today. However, it is most important that proven materials and techniques are used for any new application, or failing this, that extensive testing and production trials are carried out.

Application of composite materials to large displacement vessels is currently restricted to mine countermeasures vessels, but composites do offer considerable benefits compared with steel and aluminium for other ship types as a result of experience with MCMVs. One future major application is likely to be for superstructures and masts of larger warships such as frigates. This could give advantages of top weight-saving, resistance to fatigue cracking, improved fire performance, ballistic resistance (using higher modulus materials such as Kevlar) reduced through-life costs and reduced signatures through the use of built-in radar absorbent materials. The construction philosophy for large superstructures is likely to include steel transverse portal frames 'clad' with load-bearing FRP panels. Such an arrangement aims to minimise production and repair costs and the presence of ductile steel framing results in good resistance to nuclear air-blast effects. Special consideration must however be given to the support of weapons and trackers, which have a high support stiffness requirement, and to the problem of EMI screening.

It may not be cost-effective to use composite materials for primary hull structure in vessels above around 70 metres in length, because of the requirement for high modulus materials to resist significant bending moments. However, FRP structures could be used internally for parts of the hull such as lower decks, watertight bulkheads and minor bulkheads, where useful weight-savings could be made.

E-glass reinforcements are likely to dominate in ship construction, with higher modulus materials being applied in small quantities for special purposes. Traditionally used plain weave woven roving and chopped strand mat are likely to give way to newer fabrics such as twill, multi-axial and knitted fabrics. Resin systems are likely to be polyester or vinyl ester for external structure. Phenolic and modified acrylic resin offer superior fire-resistance and may see application internally.

2.6 REFERENCES

1] Harrhy, J., "Structural Design of Single Skin Glass Reinforced Plastic Ships", Proc. Intl. Symp. *GRP Ship Construction*, RINA, London, October 1972.

2] Dodkins, A.R., "Structural Design of the Single Role Minehunter", Proc. Intl. Symp. *Mine Warfare Vessels and Systems*, RINA, London, June 1989.

3] Smith, C.S., "Design of Marine Structures in Composite Materials", Elsevier Applied Science, London, 1990.

4] Report of Cttee. V.8, *Composite Structures*, Proc. 10th Intl. Ship Struct. Cong., Lyngby, 1988.

5] Report of Cttee. V.8, *Composite Structures*, Proc. 11th Intl. Ship Struct. Cong., Wuxi, 1991.

6] Smith, C.S., "Buckling Problems in the Design of Glass Reinforced Plastic Ships", J. Ship Res., 1972.

7] Smith, C.S., "Structural Design of Longitudinally Corrugated Ship Hulls", Proc. Intl. Conf. *Advances in Marine Structures*, DRA, Dunfermline, May 1986.

8] Smith, C.S., Chalmers, D.W., "Design of Ship Superstructures in Fibre Reinforced Plastics", Trans. RINA, **129**, 1987.

3 DESIGN OF DYNAMICALLY SUPPORTED CRAFT

3.1 INTRODUCTION

The design case study for dynamically supported craft is based on a high speed Surface Effective Ship (SES), designed and built at Karlskronavarvet, where stealth techniques have been adopted.

The vessel was built for the Royal Swedish Navy as a test vehicle with the main purpose to develop stealth techniques and to optimise the FRP structure for a military SES.

It was therefore decided that the vessel should go through an extensive test programme in various sea conditions, where structural measurements should be carried out in order to obtain experience of the dynamic loads which act on a SES at high speed in rough sea.

3.2 CONCEPTUAL DESIGN

3.2.1 General

The conceptual design approach for this vessel gave very special requirements with extremely sharp corners, specific angles between surfaces and close surface tolerances. New materials, like vinyl ester, were to be used in full scale production of a larger vessel for the first time in order to get long-term experience with this type of structure.

The vessel was to be designed with large openings for missiles in the side shell and upper deck just above the hatches for the main engines, a fact that complicated the structural design, since the openings had to be very stiff due to the stealth requirements.

The main dimensions for the vessel is:

Length between perpendiculars	L_{pp}	= 27.0 m
Breadth (max)	B_{max}	= 11.4 m
Displacement	Δ	= 130 tonnes

The section through the vessel at midships is shown in figure 3.1.

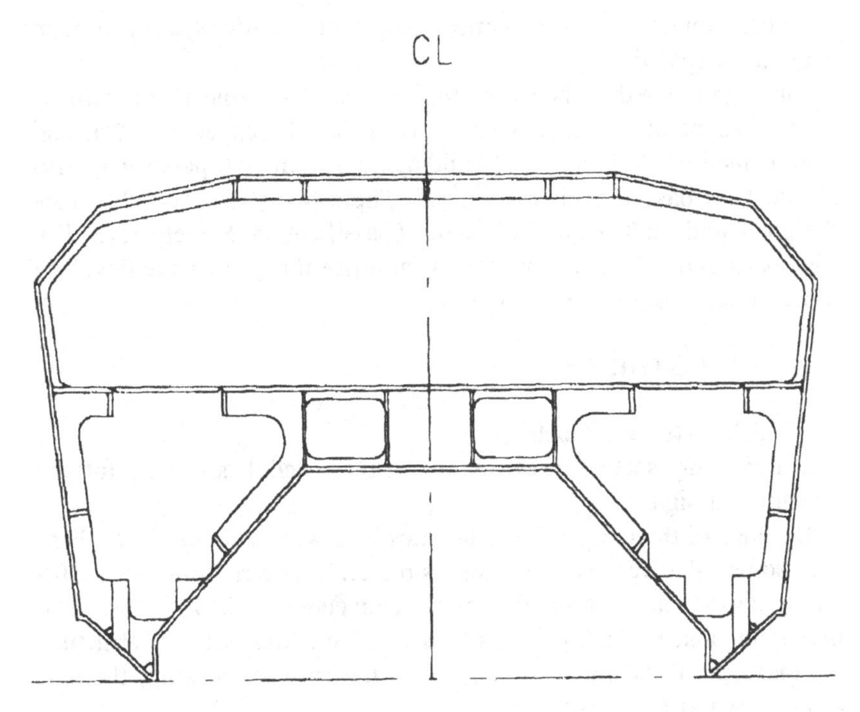

Figure 3.1. Typical section amidships.

3.2.2 Requirements

The vessel had to be optimised regarding structural hull weight and still had to have the strength to withstand dynamic loads from mine explosions. It was therefore decided that the vessel should be built as a complete sandwich design.

Since the speed requirement was in the region of 50 knots, special considerations had to be taken to avoid secondary damage from collisions with floating objects. A collision with a floating object at high speed may cause small primary damage in the bow region. This small damage can propagate to very large secondary damage where the outer face laminate can be peeled off if the bow area is not properly designed or if the core material has too low strength properties.

The noise emission to the water had to be minimised, which means that the supporting structure for the main engines had to be as stiff as possible.

The vessel had to be designed with large openings midships for the missiles, which implied that the longitudinal bending moment and torsional moment had to be carried only by the girder system in main deck and upper deck.

Since the Swedish Navy at the start of the project had limited experience of SES design it was agreed that the experience that had been gained by designing and building two 33 m, 300 passenger, SES should be a base for this project [1]. These two passenger SES were designed and built according to the Classification Society rules [2]. The loads and criteria for minimum laminate thickness were therefore to be in accordance with their rules.

3.3 LOADINGS

3.3.1 General Loadings

The following section presents the criteria and loads used for the structural design.

Because of the unique characteristics of a wide side hull SES, there are some differences between structural criteria for the SES configuration and those for more conventional hull forms. The maximum loads which principally affect the craft's primary structure are those produced by slamming when the craft is operating through a rough sea at high speed.

The maximum hull panel design pressures have a significant effect on weight since the bottom panels of the craft cover a large surface area and therefore constitute a major portion of the structural weight.

Knowledge of the speed versus sea state envelopes that ought to be used to develop loads on large SES constructed in FRP sandwich, is limited.

In general following loads have to be considered:

- Concentrated loads due to impact or accelerations
- Slamming loads acting on the hull bottom panels
- Static loads from sea water pressure
- Overall loads corresponding to general flexing of the vessel when it is in a hollow of a wave, on the crest of the wave or hits the wave in high speed
- Complex dynamic pressure load from mine explosions

3.3.2 Vertical Accelerations

The vertical accelerations along the craft's vertical axis can, according to DNV rules, be expressed as

$$a_v = (k_v g_o)/3458\,(H_s/B_{wl} + 0.084).\tau/4\,(50 - \beta_{cg}).(v/\sqrt{L})^2\,LB_{wl}^2/\Delta \tag{3.1}$$

where

k_v = longitudinal distribution factor taken from figure 3.2
H_s = significant wave height in m
β_{cg} = deadrise angle in LCG in degrees (min. 10°, max. 30°)
τ = trim angle in degree (min. 4°, max. 7°)
B_{wl} = Total breadth of thc hull
L = Craft length
g_o = standard acceleration of gravity
v = maximum speed in knots
Δ = displacements in tonnes

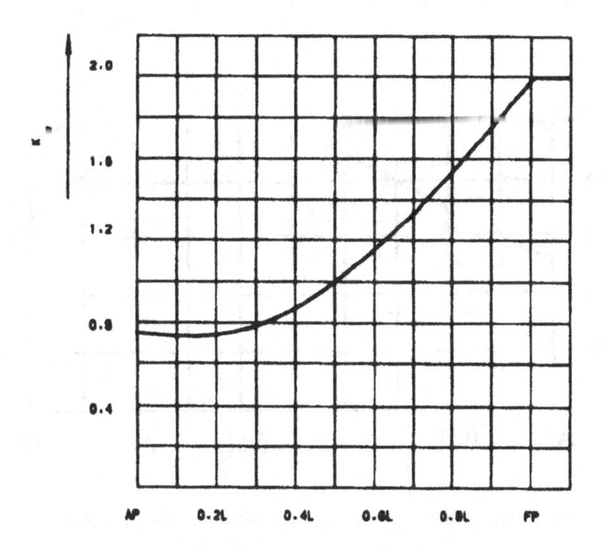

Figure 3.2. Longitudinal distribution factor for vertical acceleration.

For this particular craft the vertical acceleration is calculated for a speed of 50 knots and a significant wave height of 1.0 m.

The vertical acceleration in the bow of the vessel is then calculated to be 85.2 m/s^2 (8.7 g) and the acceleration midships in LCG is about 42 m/s^2 (4.3 g).

3.3.3 Slamming Pressure

The bottom structure and flat wet deck structure must be designed to resist the effect of slamming.

The design slamming pressure can, according to DNV rules, be written as:

$$p_{sl} = 1.3 k_l \ (\Delta/A)^{0.3} T^{0.7} \ (50 - \beta_x/50 - \beta_{cg}) a_{cg} \ \text{kN/m}^2 \qquad (3.2)$$

where

k_l = longitudinal distribution factor from figure 3.3
A = design load area for element considered in m^2
T = draught at L/2 in m
β_x = deadrise angle in degrees at the middle of the load area
β_{cg} = deadrise angle in degrees at LCG (min. 10°, max. 30°)
a_{cg} = design vertical acceleration at LCG

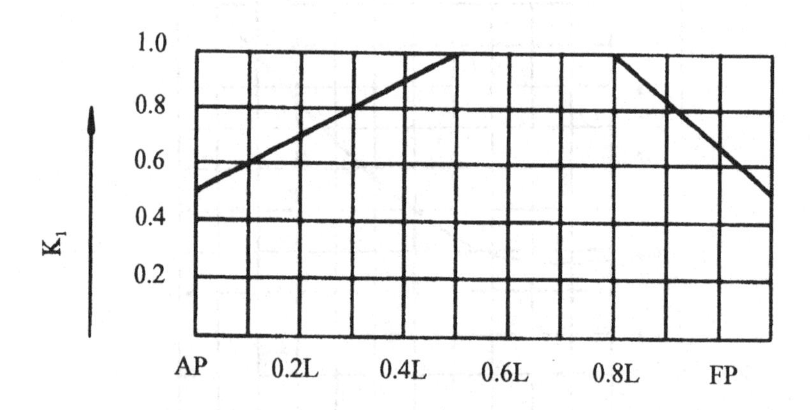

Figure 3.3. Longitudinal slamming pressure distribution factor for bottom panels.

The slamming loads that act on the dead rise area at 0.8L of the actual craft, which is to have panel sizes in that area of (0.9 x 2.3 m) 12 m^2, can then be expressed as

$$p_{sl} = 1.31 \times 0(130/2)^{0.3} \times 0.7^{0.7} \times (50\text{-}30/50\text{-}30) \times 43.2 \ \text{kN/m}^2 \quad (3.3)$$

and,

$$p_{sl} = 153 \ \text{kN/m}^2 \ (\text{or } 0.153 \ \text{MPa}). \qquad (3.4)$$

Because the design process is iterative, criteria and loads have to be reviewed and perhaps revised after the final panel sizes have been chosen. The different areas of the craft's proposed bottom structure have to be reviewed in a similar way to the example above in order to specify the slamming loads that act on different parts of the structure.

Other specific areas that have to be considered for an SES are the flat wet deck structure and the bow ramp structure between the two side hulls. Thc bow structure will probably encounter the most severe bow impact loads when the craft is running at high speed in a rough following sea during a broach when the speed is more or less momentarily reduced and the craft is diving into the wave.

The slamming pressure acting on the flat wet deck structure is, according to DNV rules, to be taken as

$$p_{sl}=1.9\, k_t\, L/A^{0.3}\ ((0.19+0.39\sqrt{L})/C_b)^2\ ((1-(H_c/0.7L))\ kN/m^2 \qquad (3.5)$$

where

x = distance from AP to load point
H_c = vertical distance from Wl to loadpoint
k_t = longitudinal pressure distribution factor taken from figure 3.4
C_b = block coefficient

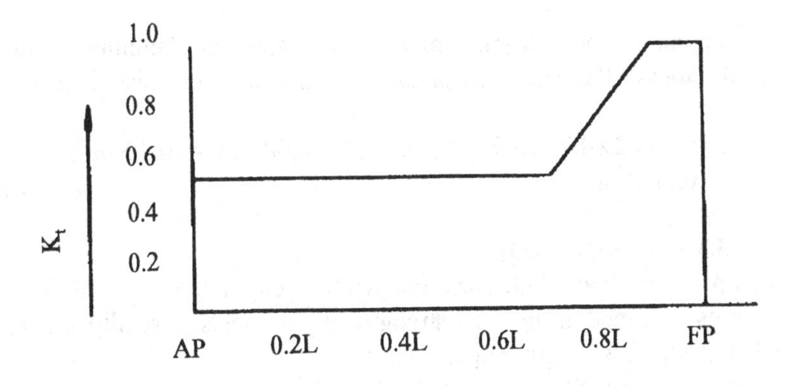

Figure 3.4. Flat cross slamming distribution factor.

For this particular ship the slamming load in the bow ramp area can be calculated to be 144 kN/m^2.

3.3.4 Static Loads

The static sea pressure that acts on the craft's side including superstructure and weather decks shall, according to DNV rules, be taken as

$$p = 10h_o + (k_s - 1.5h_o/T)\, 0.08L \text{ kN/m}^2 \tag{3.6}$$

for a loadpoint below waterline and

$$p = ak_s\,(cL - 0.53h_o) \text{ kN/m}^2 \tag{3.7}$$

for a loadpoint above the waterline where,

p_{min} = 6.5 kN/m^2 for hull side and
p_{min} = 5.0 for weather decks
k_s = 7.5 aft of midships
k_s = 5/Cb forward of FP and k_s is linearly varying between FP and midships
h_o = vertical distance from the water line at draught T to the load point
a = 1.0 for craft's sides and open freeboard deck
a = 0.8 for weather deck above freeboard deck
c = Factor depending on the service restrictions, which are minimum for this case, 0.08

For the panel in the bottom, which is calculated for slamming in the example above, the static sea pressure load can be calculated to be:

$$p = 10\text{x}2\text{x}0.7 + (0.9 - 1.5\text{x}0.7/0.7)\text{x}0.08\text{x}27 = 36 \text{ kN/m}^2 \; (0.036 \text{ MPa}). \tag{3.8}$$

3.3.5 Overall Loads

For ships of ordinary hull form and with a length less than 40-50 m, scantlings obtained from local strength requirements normally satisfy the longitudinal strength requirements.

However, for a SES concept the hull shape often is such that the longitudinal strength must be checked.

The DNV rules describe which longitudinal and torsional moment which shall be applied on the hull girder beam.

3.4 STRUCTURAL SYNTHESIS

3.4.1 General

Having decided upon foam sandwich construction for the vessel it was proposed that the complete hull should be built on a male mould. The production sequence would involve first fitting the foam core onto the mould and gluing the core together with structural putty, followed by laying-up the outer FRP laminate onto the core. Subsequently the hull would be turned, the moulds removed and the inner FRP laminate laid-up.

Through a combination of weight-savings and consideration of the ease of production, a core thickness of 60 mm was chosen for the entire hull.

A choice of a thinner core would have resulted in smaller panel sizes, which would have increased the amount of supporting stiffeners like girders and web frames, leading to increased production time and also a heavier structure since a minimum laminate thickness for robustness reasons must be obtained despite low panel stresses.

The material properties that were to be used for the scantling analysis were based on the extensive testing of all materials that were carried out by the Swedish Material Administration prior to the design start of this particular project in order to optimise the material combinations.

Many different glass fibre combinations were tested with different types of resins, like urethane modified isophthalic polyester resin, different types of vinyl esters etc. Special attention was given to the delamination strength. The finish of the glass fibre is very important for the delamination strength of the laminate. Tensile strength, which normally sets the limits, is not so sensitive to variations of sizing of the glass fibre material as the delamination strength. Experience has shown that laminates made with glass fibre of the same type, from the same manufacturer but where sizing of the fibres was changed, had about the same tensile strength but only 40% of the predicted delamination strength.

Other tests that were carried out were impact tests with different types of fibres and resins. From these tests it was decided that the laminates should mainly be built-up with combinations of conventional E-glass fibres consisting of a combination of 800 g/m^2 woven rowing and 100 g/m^2 chopped strand mat, CSM. For attachment lamination of bulkheads, pre-cut glass fibre with a different number of plies in ± 45° fibre direction were to be used.

The matrix material for the hull panels exposed to dynamic loads from slamming and mine explosions as well as for the attachment laminates of bulkheads was to be made with a rubber modified vinyl ester resin. The resin used for the rest of the vessel was to be a standard isophthalic polyester resin.

The minimum requirements for tensile and compressive strengths that the laminates were to fulfil were set to 180 MPa in 0°and 90° fibre directions of the woven roving with a Young's modulus of 12000 MPa. The target for the delamination strength was set to be about 17-22 MPa for the laminates built-up with vinyl ester resin and 10-15 MPa for laminates built-up with normal isophthalic polyester resin.

The core material was to fulfil several different requirements for an optimised sandwich design like:

- low density
- high shear strength and stiffness
- high shear strain at failure

The correct specification of the core material was therefore essential and the different qualities had to be chosen carefully. Since the interaction of the laminate and core is so important in a sandwich design, there is, of course, a strong relationship between the characteristics of the core material and the laminate.

The effect of too low core quality is especially hazardous in the bottom of a fast vessel which is subjected to high impact loads caused by slamming.

Too low a core quality may cause shear fatigue due to the slamming loads after a relatively short time, but also cause severe secondary failures such as laminate peeling due to hydraulic pressure caused by collision with floating objects.

The core material to be used for the vessel was chosen to be rigid PVC foam with different densities for the different parts of the hull structure, varying from 60 to 200 kg/m^3. The core material was to be glued together to form a homogenous core structure. For this purpose a structural polyester putty is normally used. The mechanical strength for a polyester putty will normally satisfy the strength requirement which is the same as for the adjacent core material. However, normal polyester putty has a very low elongation to failure, about 1-2%

compared to the core which normally has an elongation to failure of about 6-10%. For impact loaded structures it is therefore essential that the putty used for joining the core material together has similar properties as the core material.

The best available low weight polyester putty on the market at the time for this project, that fulfilled the mechanical strength properties (min. 6MPa) and could be pumped in a conventional pump equipment, had an elongation to failure of 3.5%. It was, therefore, decided that this type of putty was to be used for the entire vessel.

3.5 STRUCTURAL ANALYSIS

3.5.1 Modelling Technique

The structural analysis was primarily based on the requirements of DNV rules for High Speed Light Craft [2].

Since the design process is iterative, the first approach to hull scantlings were checked against the lateral pressures that were calculated for the different parts of the vessel. Normal stresses in the laminate, shear stresses in the core and panel deflections were calculated for the different parts of structure.

Stresses

Maximum normal stresses in the skin laminates at midpoint of a sandwich panel subjected to lateral pressure is given by DNV and can be written as

$$\sigma_c = \frac{160\, pb^2}{W} C_n \qquad N/mm^2. \tag{3.9}$$

C_n = $C_2 + \nu C_3$, for stresses parallel to the longest edge, see figure 3.5.
= $C_3 + \nu C_2$, for stresses parallel to the shortest edge, see figure 3.5.

W = Section modulus of the sandwich panel per unit breadth in mm^3/mm. For a sandwich panel with skins of equal thickness $W = dt$.

The maximum core shear stresses at the midpoints of the panel edges of a sandwich panel subject to lateral pressure is given by:

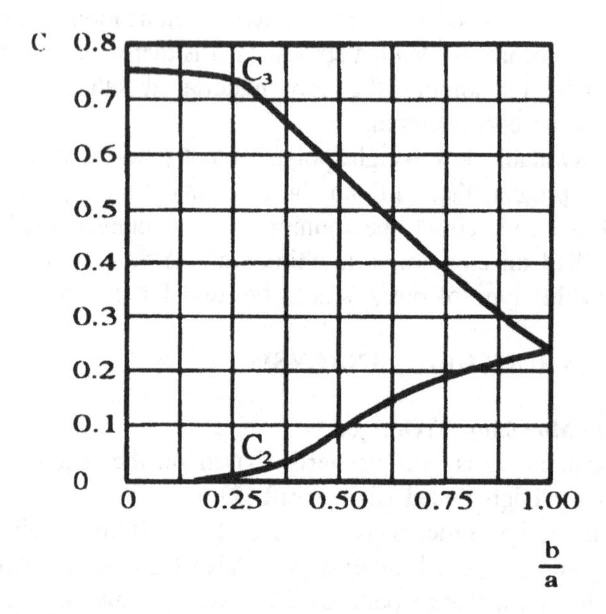

Figure 3.5. Plots to determine C_2 and C_3.

$$\tau_c = \frac{0.52\ p\ b}{d} \qquad N/mm^2. \qquad (3.10)$$

C_5 = C_4, for core shear stress at midpoint of longest panel edge, see figure 3.6.
= C_5, for core shear stress at midpoint of shortest panel edge, see figure 3.6.

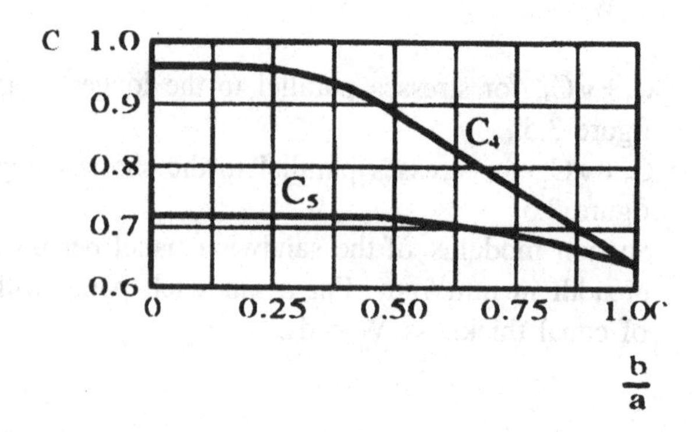

Figure 3.6. Plots to determine C_4 and C_5.

Deflections

The deflections at the midpoint of a flat panel is given by:

$$w = \frac{p\,b^4}{D_2}\,(C_6 + p\,C_7)\,10^6 \text{ mm.} \qquad (3.11)$$

For panels with skin laminates with equal thickness and modulus of elasticity:

$$D_2 = \frac{E\,t\,d^2}{2(1-\nu^2)} \qquad (3.12)$$

For panels with skin laminates with different thickness and modulus of elasticity:

$$D_2 = \frac{E_1\,E_2\,t_1\,t_2\,d^2}{(1-\nu^2)(E_1\,t_1 + E_2\,t_2)} \qquad (3.13)$$

The indices 1 and 2 denote inner and outer skins respectively.

$$p = \frac{\pi^2\,D_2}{10^6\,G\,d\,b^2} \qquad (3.14)$$

C_6 and C_7 can be obtained from figure 3.7.

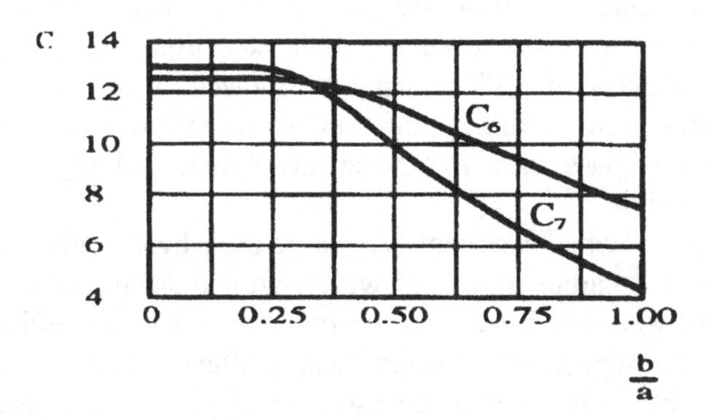

Figure 3.7. Plots to determine C_6 and C_7.

Using the panel in the deadrise area at 0.8L as an example for the analysis, it can be seen that the maximum load on the panel is caused by the slamming pressure p = 0.153 MPa.

The deadrise area had the following conceptual dimensions:

Panel size a = 2.3 x b = 0.9 m
Laminate thickness 5.0 mm with ultimate tensile stress σ_u = 180 MPa.
Core thickness 60 mm with ultimate shear stress τ_u = 2.9 MPa.

By using the equations above, the following stresses in the laminate and core were obtained:

Laminate stress σ = 40.10 MPa. ($\simeq 0.22\sigma_u$)
Shear stress in the core τ = 1.0 MPa. ($\simeq 0.35\tau_u$)
Panel deflection at midpoint δ = 11 mm
Deflection ratio δ/b = 11/900 $\simeq$ 0.012

This fulfils the requirements for the actual panel.

3.5.2 Constraints

Constraints were put on the maximum stresses occurring in the laminate and the core and on deformations of the hull panels. All constraints were based according to DNV rules except for areas where specific stiffness requirements were needed, such as the supporting structure for the 40 mm gun of Trinity type which required extremely stiff foundations in order to achieve the performance requirements.

A finite element analysis using about 300, 8-node isoparametric, sandwich elements was carried out for the supporting structure around the gun foundation. This analysis showed that the stiffness requirements were extremely difficult to achieve with conventional unidirectional glass fibre reinforcements without adding too much structural weight.

A possibility which was analysed was to use a hybrid-reinforcement with glass and carbon fibres. However, an additional finite element analysis showed that it was much more cost and weight-effective to reinforce the supporting structure with stiffeners mainly made of carbon fibres. Full scale measurement after the final construction showed a very good correlation between calculated and measured stiffness.

Another area with special stiffness requirements was the supporting structure for the main engines due to the requirements of noise emission to the water. A finite element analysis for this particular structure showed that the stiffness and noise requirements could be fulfilled in a weight-effective manner by using a hybrid laminate with glass and aramid fibres in this area.

3.6 PRACTICAL CONSIDERATIONS

3.6.1 Production Features

Throughout the design, production aspects were of prime concern and feedback from production was continuously given to the designers. The weight requirements for the entire vessel were severe since a high speed SES is very sensitive to weight, and a strict weight control of the hull with feedback both to production and design office was needed in order to be able to take corrective actions or, as a last alternative, to redesign certain equipment.

During the construction of a large reinforced composite structure, it is essential that the design is such that the laminating and bonding processes are easily carried out, otherwise the required design properties will not be obtained. The production engineering approach was to use the well proven building methods developed for the larger Mine Counter Measurement Vessels built at the Yard according to the GRP sandwich concepts.

The building technique using simple wooden male moulds and a foam sandwich design is extremely competitive for the construction of different crafts in small series.

The advantage of this building method compared to the method using female moulds is the low cost of the mould compared to the much higher cost for a female mould. Another important advantage of this construction method is the simplicity of ensuring good bonding between the core and the outer laminate. This can be more difficult to control in large structures using female moulds and vacuum bagging technique. A disadvantage of the male mould method is of course the rougher outer surface, which requires extra sanding and putty work if the surface finish requirements are high.

At an early stage, there was discussion regarding whether or not to divide the shell into sections and build the two side hulls separately and to produce the flat wet deck as prefabricated flat panels, which would be joined later to the two side hulls. With this type of arrangement the turning over procedure of the hull after laminating the outer laminate would be simplified.

However, with a tight time schedule and the special stealth requirements, this method was regarded as a potential risk and the hull was therefore divided along the ship centre line and produced as two units which were later laminated together.

Flat panel production is the most production friendly and productive manufacturing process when using composite materials. Problems with thixotropy of the resins, which are very common when using vinyl ester and low weight polyester putty for bonding core planks together, are minimised with this type of production. Bulkheads and web frames were manufactured as flat panels. Large flat panels such as decks were produced in sections with both sides fully laminated prior joining them together.

Water jet inlets, which have an extremely complicated shape and demands of surface finish were laminated on female moulds. The core material around the inlets was glued into position after the inlet was laminated to the hull.

3.6.2 Cost Constraints

It has been a general belief that ships built of FRP are considerably more expensive than corresponding vessels built of steel or aluminum. However, experiences gained from building similar ships for more than 20 years clearly shows that this is not the case. It is true that the material cost for the hull itself is higher, but the total working hours for building and outfitting the complete ship is lower.

For fast vessels like the passenger SES, which have many running hours per year, the weight-saving by using more expensive materials that give better strength properties, together with a structural optimisation, can be well worth the additional costs. A structural optimisation of a 33 m passenger SES [3] using a computer optimisation program, showed that it was possible to reduce the structural hull weight up to 30% if only stress and deformation criteria were used. Of course it is not possible to reduce the hull weight to that extent, since demands of robustness and other practical aspects require a more uniform design, but for a passenger SES, where operational cost savings are of greatest interest, the potential weight-saving is very useful and can give a good pay-off despite higher production cost.

The principal operating costs are related to fuel costs, maintenance, repair and service life. Significant fuel savings can be obtained if the reduction in structural weight allows smaller engines. Maintenance of the structure involves inspection and painting. Inspection of a FRP structure is more difficult than for steel or aluminum structures, since FRP procedures and technology still are under development, but are firmly established for steel and aluminum. Repair of FRP structures require good environmental conditions, but the procedures are relatively straightforward. There are a number of guidelines for repairing FRP and FRP sandwich structures, but they are not standardised in the same way as for steel and aluminum, due to the wide variety of material combinations available and continuing development of the material. Small field repairs can be handled, with repair kits by trained personnel, quickly and easily. Major repairs require shore-based facilities experienced in dealing with FRP.

Service life for a correctly designed and built ship is potentially very long if necessary maintenance like painting is carried out and if the UV degradation of the laminate is avoided. Experiences from well documented ships with almost 20 years in service, show no mechanical degradation of the hull structure where test samples have been taken.

3.6.3 Legislative Features

One of the most important safety considerations is fire resistance. The important aspects of fire-resistance are the minimisation of toxic gases, prevention of the spread of fire to adjacent compartments and the maintenance of the structural integrity.

The fire resistance of a FRP-sandwich design using fire retardant additives or intumescent coatings gives better fire-resistance than aluminum. However, as long as there are no international general regulations for FRP designed vessels regarding fire protection, similar to steel and aluminum, the approval of the FRP design must be carried out by domestic authorities. This can be a problem if later the vessel should be sold to another country with a different domestic philosophy regarding fire protection and approval of the design.

Another legislative aspect regarding construction of FRP products are the different domestic laws and regulations regarding styrene emission to the workshop and surrounding atmosphere which leads to different economical conditions in different countries.

3.7 CONCLUSION

It is clear that FRP is a superior material for construction of large vessels and small crafts. Most of the disadvantages occur from incomplete knowledge of the material behaviour and lack of experience of the material.

For extremely fast vessels the limited knowledge of the load spectra on which the vessel should be designed for have caused unnecessary failures. In view of the extensive research work being carried out by

Classification Societies and navies, these problems will certainly be avoided in the future and there is a tremendous potential for future increases in the performance of FRP structures.

3.8 REFERENCES

1] Hellbratt, S-E., Gullberg, O., "The High Speed Passenger Ferry Jet Rider" Proc. 2nd Intl. Conf. *Marine Applications of composite Materials*, Florida Institute of Technology, Melbourne, March 1988.

2] "Tentative Rules for Classification High Speed Craft, Det Noreske Veritas, Hovik, 1990.

3] Romell, G.O., Ljunggren, L.A., Esping, B.J.D., Holm, D., "Structural Optimization of a Surface Effective Ship", Proc. 1st Intl. Conf. *Sandwich Constructions*, Stockholm, June 1989.

4 THE ROLE OF ADHESIVES

4.1 INTRODUCTION

Composite materials owe their very existence to adhesion between resin matrix and reinforcement. However, despite the general applicability of much of this Chapter to this aspect of adhesion, the intention is to focus interest on the more demanding problem of 'structural' adhesion as an alternative means of joining composite components together to facilitate ease of fabrication from standard components. A number of excellent general guides, [1,2] as well as more detailed texts [3,4,5], cover the basic principles and these have been drawn on, and supplemented where appropriate by reference to recent research papers.

4.1.1 Nature of Polymeric Adhesion

Although adhesives of animal, vegetable or mineral origin have been in use by man for millennia, it has only been in the last 50 years that the development of synthetic resins have accelerated to a point where there is now a bewildering array of adhesives available, many of sufficient potential strength to be labelled 'structural', compatible with almost all possible adherents. Choosing the right adhesive depends on chemical, design and fabrication considerations grafted onto a basic understanding of the nature of adhesion itself.

Bonding depends for its success on the forces which exist at the molecular level between the adherent and adhesive and within the adhesive itself. It follows that to achieve a successful bond requires that:

a) the molecules of the adhesive come into intimate contact with those in the surface of the adherent
b) this adhesive contact is not lost for any reason over time
c) the cohesive strength of the adhesive material is adequate to sustain the mechanical strains imposed on the joint. This should include not only consideration of static and dynamic loading but also temperature and creep properties

Condition (a) can only be achieved if the adhesive thoroughly wets the adherent surfaces. This process will be inhibited by the presence of oily films or release agents which the adhesive cannot completely absorb. Condition (b) requires that undesirable molecules do not migrate to the adhesive/adherent interface to displace adhesive molecules while the final condition depends on the cohesive behaviour of the adhesive between adherents, a function of such adhesive properties as modulus, toughness, and glass transition temperature (Tg).

In general, adhesives must undergo a non-reversible change from a liquid or viscous phase which wets the adherents before curing to a solid. The process by which this happens leads to a simple classification [1]:

1. solvent based - solidifying through the loss of a solvent
2. temperature-setting (thermoplastic) adhesives - liquified by heat and solidifying on cooling
3. chemical-setting - relying on chemical reaction for their cure (with or without heat)

Class 3 includes most of the elastomers and thermosets which find uses as either the matrix of FRP composites or structural adhesives. Unlike thermoplastics, they are essentially infusible and insoluble but may be reformulated to enhance particular properties. Thus the number of products on the market continues to grow.

4.1.2 Aspects of the Adhesive Bonding Process

Apart from understanding the general strengths and weaknesses of adhesives, which lead to a number of design considerations discussed in the next section, the major task is to narrow down suitable generic families of adhesives from which to make a selection based on performance, application and cure parameters.

4.1.2.1 Matching adhesive to adherents

References [1] to [5] all give tables of suitable generic adhesive types to bond most combinations of adherents. However, in all cases designers are advised to seek technical assistance from manufactures before final decisions are made. When dealing with FRP materials (and particularly their possible combination with alloy and steel components) Lees [7] reduces the choice to:

- epoxy: one and two part
- acrylic: true one part, pseudo one part and two part
- polyurethane (PU): one and two part

In addition, use of the FRP base resin might be considered, but this will not generally offer the same range of capabilities as the first two, particularly in structural joints. Where structural joints are required then only the stiff (high modulus) epoxy and acrylic formulations offer strong durable joints. However, stiff adhesives are also brittle and to overcome this both of these families have 'toughened' versions which incorporate a microscopic rubbery phase dispersed throughout the matrix providing internal crack-arrestors to inhibit crack propagation.

4.1.2.2 Adhesive application and cure

To obtain durable structural joints some surface preparation will always be necessary. Application of the adhesive, usually to one adherent surface, requires some care to ensure that air is not trapped. This is best accomplished by a single continuous bead applied directly from manual or automatic dispensing equipment, using a weave pattern where necessary on wider joints or where significant gap filling is required. Most of the families outlined above have good gap-filling capabilities, being thixotropic in nature, an important feature in joints involving the rough surface of hand-laid FRP materials. Automatic mixing within the dispenser nozzle is now recommended for most two part adhesives to ensure both correct mixing ratios and dispersion of the components without air entrapment.

After application there will be a variable time in which the components must be brought together. In general this is related to the gel time and will vary on formulation from minutes for acrylics to an hour or so for some two part epoxies at room temperature. Following this there will be a further working time in which manipulation of the joint is still possible, although care should be taken not to break the film of adhesive between components which will result in air entrapment. A period of up to 24 hours is then required in which the adhesive will gain sufficient strength for handling during which the structure must be securely clamped. In general, the stiff structural adhesives exhibit very small shrinkage on cure and therefore do not require high clamping pressure nor do the resulting joints suffer significant built in stress. Most joints will also benefit from the sealing and stress modifying effects of the spew fillet formed by expulsion of excess adhesive from the joint which may be smoothed as it gels, but should not be removed.

In general, raising the temperature of the joint above ambient, will accelerate the curing process and result in stronger joints through more complete cross-linking within the thermosets. Care must be taken not to overheat either adhesive or adherent and, in the case of FRP materials, this may prove difficult because of their excellent insulation properties. For this reason the single part, heat curing epoxies have only limited prospects of application in this field despite their superior performance characteristics, particularly at elevated temperatures.

4.1.3 Advantages and Limitations of Adhesives

Depending on the application, adhesives offer a number of possible advantages over the use of bolts and rivets in fabricating FRP composite structures:

- improved surface finish and fewer irregular surface contours
- can accommodate complex shapes
- more uniform stress distribution in joints
- less local damage to component materials
- good fatigue and damping capabilities
- self sealing with good insulation properties
- possibilities for simplified, self-jigging assembly

Some limitations do exist. In many cases limited numbers of bolts may be retained to accommodate extreme conditions or facilitate ease of fabrication and handling. In particular, care should be taken to evaluate some or all of the following possible problems:

- dependence of durability on processing conditions
- the need to adapt joint designs to avoid cleavage failure
- limited resistance to extreme service conditions - particularly heat and shock loads
- may creep at high temperatures
- optimum strength is not realised immediately
- specialised jigs may be required for assembly
- bonded structures are not easily dismantled for repair
- no straightforward, non-destructive quality control procedures exist. Reliance must be placed on rigorous process control,

sample joint testing and visual techniques.

4.2 DESIGN CONSIDERATIONS

The five main types of load which bonds may be required to resist are compressive, tensile, shear, cleavage and peel. These are illustrated in figure 4.1. Unfortunately adhesives do not respond in the same way to each of these and it is rare for practical joints to fit into any one of these clearly defined regimes. In general, adhesives cope well with compressive and shear stresses, but because of joint distortion and non-uniform stress distributions are generally very susceptible to cleavage effects. These have a marked effect on tensile performance as it is difficult to guarantee symmetrical tensile forces without cleavage components which will dominate joint performance if present. These points are illustrated by figure 4.2 which shows the strained shape (exaggerated) of a standard lap-shear joint and its associated stress distribution.

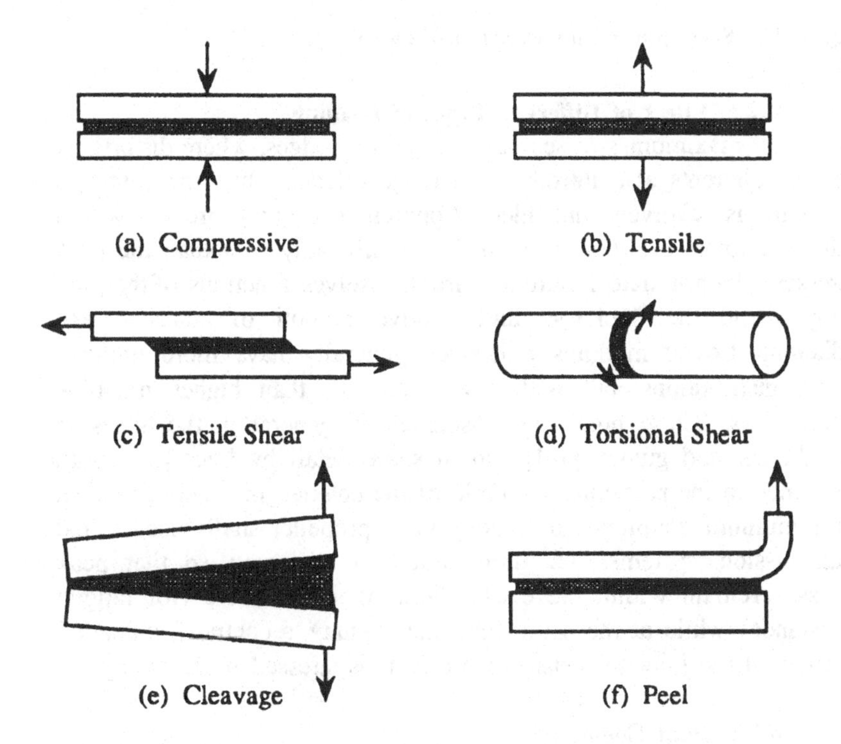

Figure 4.1. Types of load.

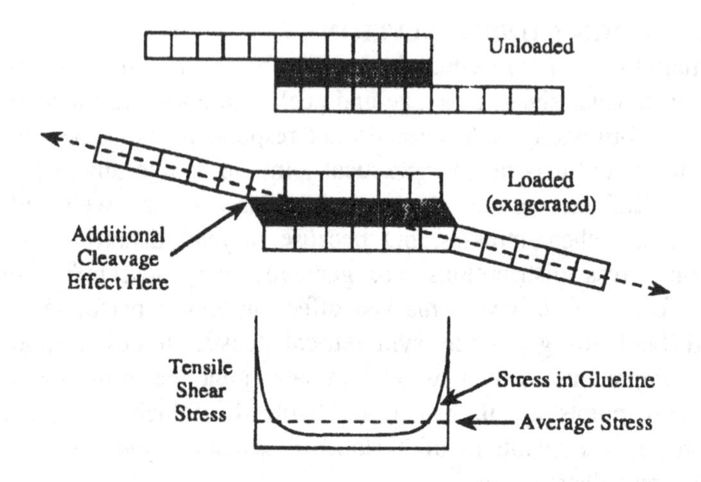

Figure 4.2. Stress pattern on a loaded lap-shear joint.

4.2.1 Effect of Different Types of Loading

Note that maximum stresses occur at the free edges, where distortions of the adherents also introduce cleavage effects, while the centre of the joint is relatively unloaded. Apparent (average) stresses within adhesive joints are therefore usually significantly less than the peak stresses which initiate failure and are themselves functions of the joint length, and the thickness and relative moduli of adhesive and adherent. Lower modulus adhesives generally have more uniform stress distributions but with lower strength than higher modulus alternatives. These points are discussed in general in the adhesive handbooks and guides [1-4] and in some detail by Lees [6,7] with reference to the particular example of the coaxial joint between steel or aluminium couplings and epoxy GRP propeller shafts used in car transmission systems. The importance of designing so that peak stresses remain within the elastic limit of the adhesive (for fatigue resistance) while at the same time maintaining a centrally unloaded portion of the joint to resist creep effects is stressed by Lees [6].

4.2.2 Joint Geometry

The minimisation of cleavage and peel effects leads directly to a number of preferred joint concepts which are illustrated as DOs and DON'Ts in figure 4.3. Butt joints and corner connections are therefore

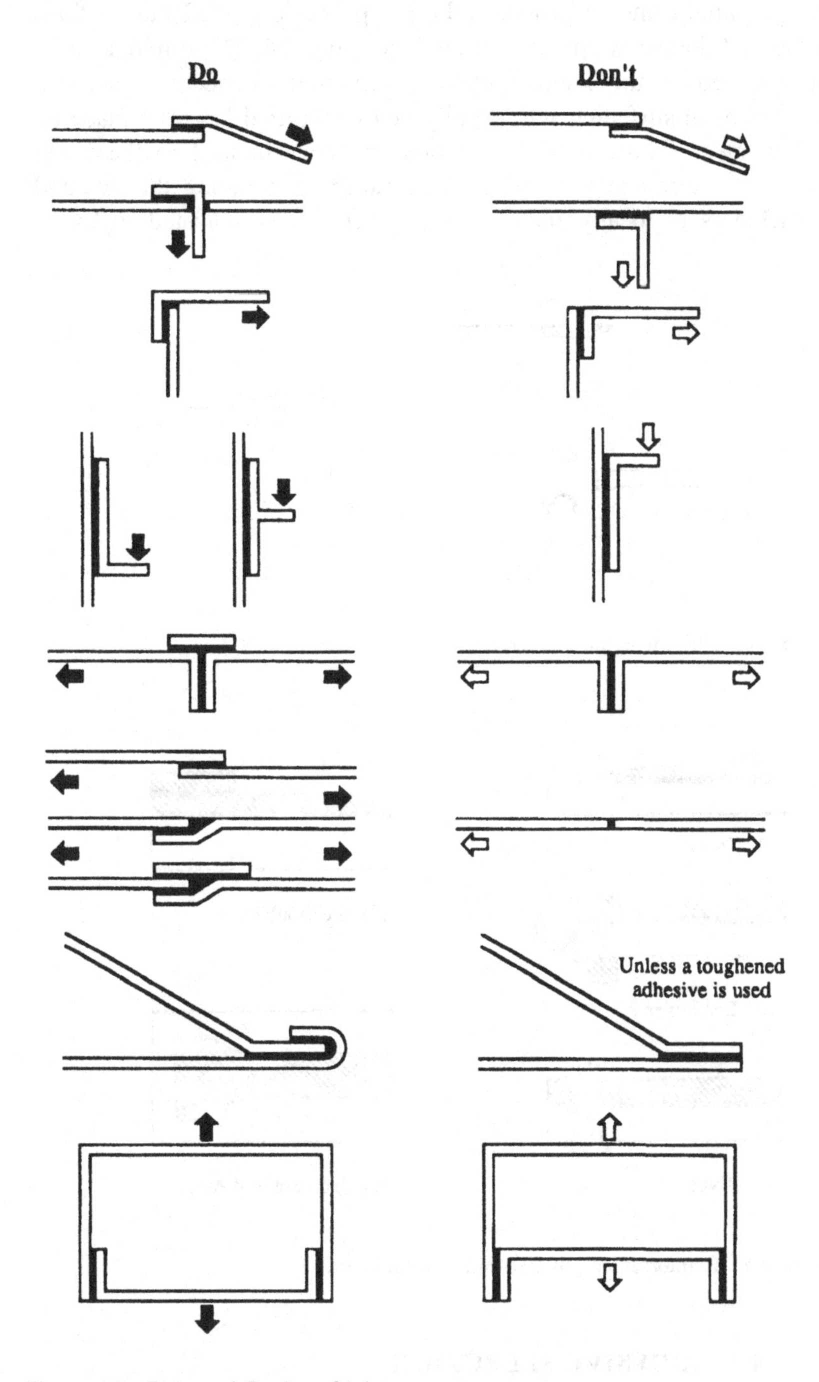

Figure 4.3. Do's and Don'ts of joint geometry.

best prepared using extrusions (alloy) or pultrusions (FRP) in the form of tapered double straps as indicated in figure 4.4. These may also be incorporated as the lower flange of stiffening members. Otherwise attachment of stiffening will usually be by modified lap-joint made up of the lower flange(s) of a stiffening member bonded to the panel surface. Special attention needs to be paid to reinforcing the free end of stiffeners or locally reducing their stiffness as indicated in figure 4.5.

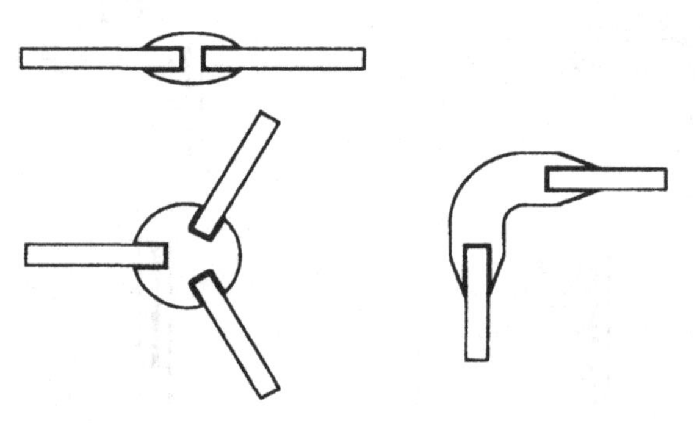

Figure 4.4. Butt joints and corner connections.

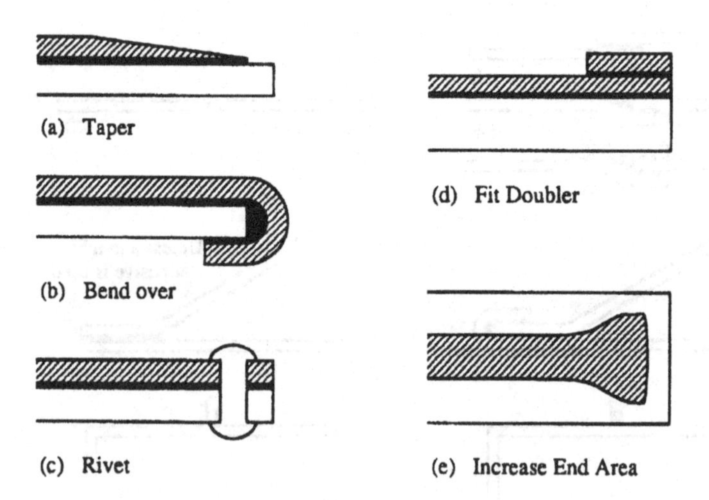

Figure 4.5. Reinforcement of free-ends of stiffeners.

4.3 ADHESIVE SELECTION

Selection of adhesives for any application should be based on thorough consideration of compatibility of materials, stress regime, joint geometry and fabrication methods proposed and the mechanical, thermal, creep and durability properties of the candidate adhesives. In this respect, it is important to conduct a range of tests relevant to the particular application on a broad spectrum of candidates before identifying the long-term solution. This implies considerable design and development lead time before decisions can be taken on fabrication methods (or components) which may have capital investment implications.

4.3.1 Adhesive/Adherent Compatibility

Initial candidate families will be chosen from the tabulated compatibility data referred to in section 4.1.1. This will produce a short list for FRP structures such as table 4.1 (suggested by Lees [7]) from which it will readily be appreciated that judgements concerning stiffness (modulus) and strength must be weighed against practical considerations such as application and cure cycles and long-term durability factors. However, the primary significance of modulus on strength leads to significantly different application potential.

At extremely low moduli (<20 MPa), materials such as one part PU's are best considered as being no more than sealants with no significant load-bearing capability.

Low to medium modulus materials (20-500 MPa) are potentially suitable for the mounting of windscreens and lightweight, non-load-bearing panels where the weight of composite involved is small relative to both the area and width (generally the perimeter of the panel) of the bonded joint. The importance of differentiating between length (parallel to applied load) and width (perpendicular) of a joint is important in this context as stress levels are generally inversely proportional to width while being little affected by length unless the adhesive is extremely ductile or the length is very short. Low modulus PU adhesives may have potential where differing thermal expansion characteristics of thin adherents might otherwise induce high local stresses or cosmetic distortion.

At the top of the modulus spectrum lie the heat cured epoxies (1000-5000 MPa) where their high potential strength required for structural applications may not always be realised because of differential contraction effects in adherents (resulting from a hot cure) or the higher edge stresses resulting from these stiff materials. As a compromise it may be necessary to consider using the lower modulus (500 MPa) two part epoxies specially formulated to achieve this end

Table 4.1. A summary of the characteristics of the principal structural adhesives for composite bonding.

	Main Characteristics	Principal Advantages	Principal Disadvantages
PU One Part	Low modulus. Very low strength.	Very simple to use. Hot metal variants very convenient on suitably sized components. Fills large gaps. No mixing	Sensitive to moisture. Not true structural adhesives. Slow curing. Must be applied to non-metal surface for long term durability.
PU Two Part	Very low to medium modulus. Very low to medium strength.	Fast curing possible. Very good application characteristics. Fills large gaps.	Sensitive to moisture. Often requires heating to achieve acceptable production times. Must be applied on a non-metallic surface for long term durability. Some versions cannot be considered to be structural adhesives. Must be mixed.
Acrylic Pseudo One Part	Medium modulus. Medium strength.	Very fast. Very easy to apply. Extremely durable. Bonds metals particularly well. A true structural adhesive. No mixing.	Needs good fit and narrow gaps to function effectively. Best below 1 mm.
Acrylic Two Part (VOX)	Medium modulus. Medium strength.	Fast. Easy to apply. Benefit of delayed action cure (DAC). Extremely durable. Fills large gaps. Copes well with light contamination. A true structural adhesive.	Must be mixed.
Epoxy One Part	High modulus. Very high strength.	Fast curing. Easy to apply. Extremely durable with robust all round performance. No mixing.	Needs to be heat cured.
Epoxy Two Part	Medium to high modulus. Medium to high strength.	Easy to apply. Durable. Can be speeded by warming/heating. True structural adhesive.	Must be mixed. Slow curing.

which are still able to withstand substantial structural load over prolonged periods [7]. In general, those toughened adhesives in the top of the acrylic stiffness range and the spectrum of cold/warm cure two part epoxies are all true structural adhesives for FRP applications.

Many of the problems discussed above only become apparent after detailed 3-D elasto-plastic finite element analysis of joint configurations for which accurate bulk modulus data for all the adhesives is unlikely to be available. In general therefore, it is impractical to undertake such studies when trying to evaluate candidate adhesives for particular applications. Instead it is normal to rely on a programme of mechanical testing on suitably chosen specimens which might also include the temperature and creep effects if these are significant to the design.

4.3.2 Mechanical Properties

Reference [10] details a number of possible small scale tests which have been used to evaluate the primary mechanical properties of adhesive/adherent combinations. Figure 4.6 illustrates the range of test specimens that were used to evaluate the relative performance of a range of candidate adhesives in terms of tensile shear, bending/cleavage and shear impact. Table 4.2 indicates the average stress at failure obtained from a representative selection of two part epoxy adhesives between steel and polyester/E-glass adherents in these tests. Figure 4.7 illustrates a further series of tests performed using one of these adhesives in which the adherents were systematically varied to produce the results in table 4.3. Notice that the average stress levels at failure are very noticeably dependent on the adherent stiffness.

Table 4.2. Comparative performance of epoxy adhesives.

Type	Supplier	Impact Energy [J]	Cleavage Force [kN]	Shear Force [kN]
2005	Ciba Geigy	38	3.7	34
2004	Ciba Geigy	23.5	2.8	30
E32	Permabond	31.5	3.8	37
E34	Permabond	19	2.8	17
9323	3M	28	3.7	35
1838	3M	27	3.8	30

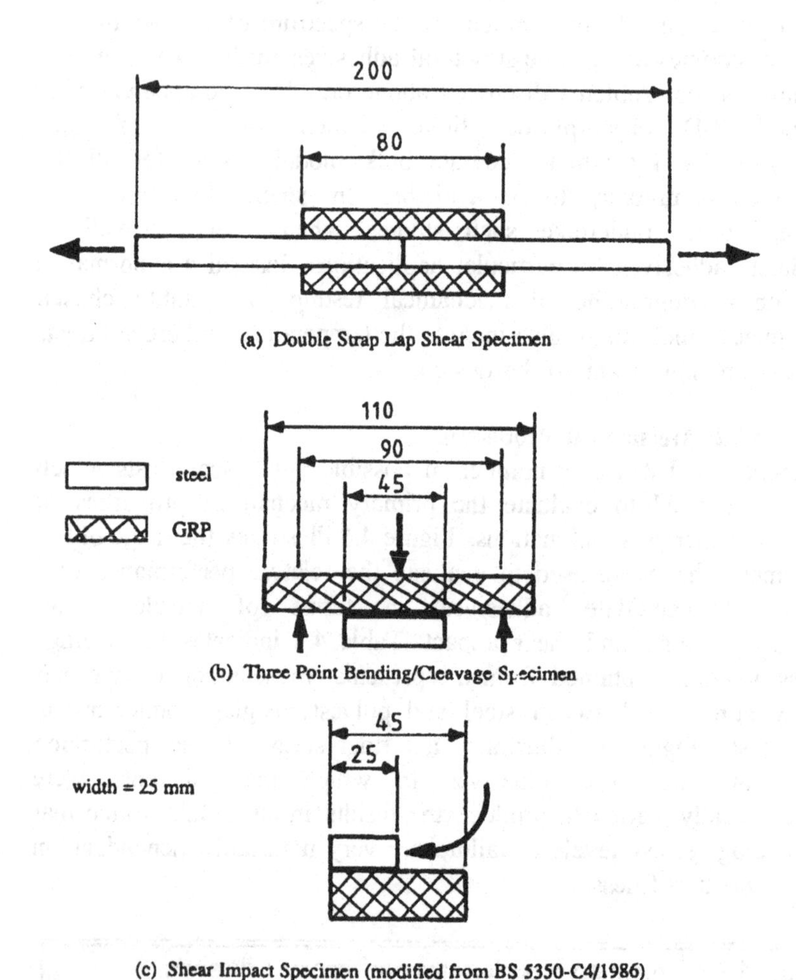

Figure 4.6. Range of test specimens.

It is not possible to infer the stress distribution in the joints at failure directly, but subsequent finite element analysis of the steel/GRP lap-shear specimen at failure load indicated tensile cleavage stresses within the adhesive 0.1 mm from the GRP surface which vary between a maximum of 150 MPa adjacent to the free edge and 10 MPa in the centre. The corresponding shear stresses are in the range 85 to 10 MPa although the average applied shear stress at failure is only

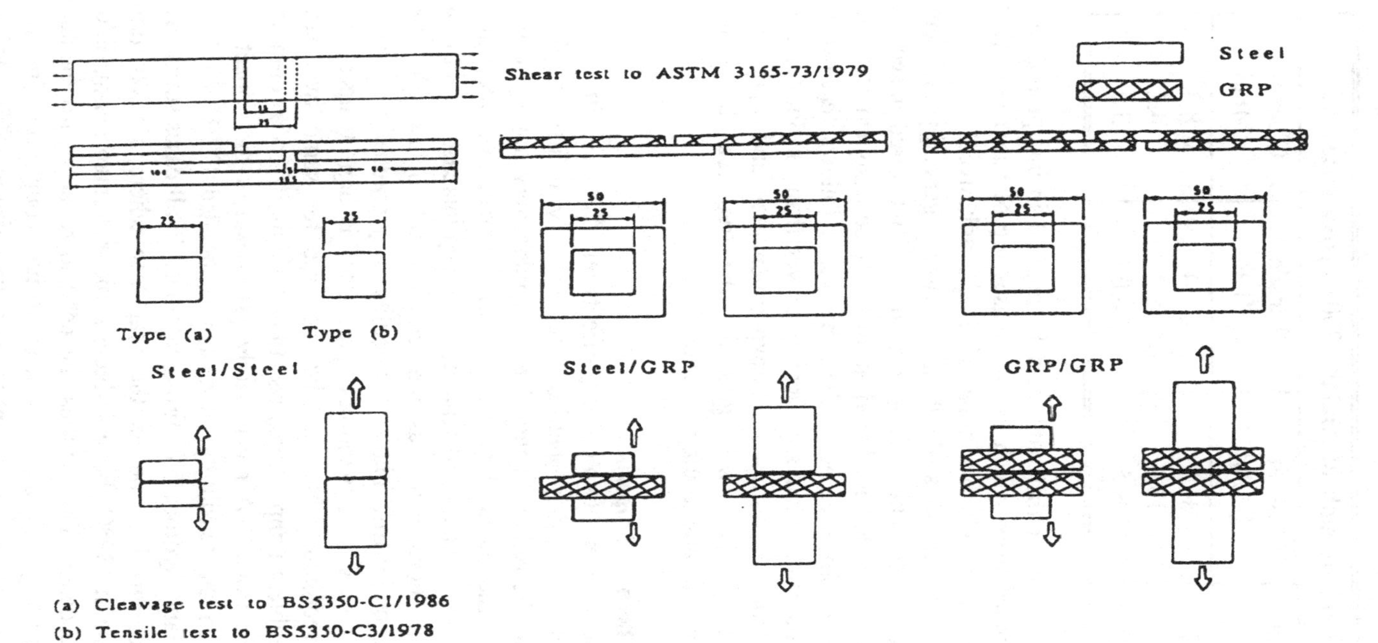

Figure 4.7. Tests on adhesives.

Table 4.3. Bond strength of Araldite 2004 epoxy adhesive.

Combination	Shear [MPa]	Tensile [MPa]	Cleavage [MPa]
steel/steel	34.0	33.0	8.3
steel/GRP	13.0	15.2	4.2
GRP/GRP	9.5	8.0	3.5

13 MPa. It is thus clear that care must be applied in interpreting the results of standard tests which in most cases are of value only as a basis for comparison within a range of similar adhesives.

The fatigue characteristics of adhesive joints are generally excellent if loading is moderate. This corresponds to the point raised in section 4.2.1 that peak stress levels should at all times be designed to cycle within the elastic stress range of the adhesive. It follows that cyclic loads should be maintained significantly lower than the limit loads indicated by static test results. Some experimentation and supporting finite element analysis will generally be required to verify performance where fatigue may be critical.

4.3.3 Thermal Properties

A rise in temperature has two primary effects on adhesives. First, their shear and tensile strength properties reduce, slowly at first, and then dramatically in the region of their Tg, remaining very low up to the point at which the material begins to char. Second, over the same temperature range their natural tendency to creep increases rapidly in similar fashion. Fortunately, as temperatures fall below 0°C the converse is generally true, although ductility may gradually reduce.

Tg is therefore a very important parameter if high temperature operation is required. This may be the case for some two part adhesives if ambient temperatures rise to as little as 50°C - a common situation in the tropics. A recent study [9] considers the impact of these effects on the feasibility of designing bonded GRP fire/blast walls to withstand hydrocarbon fire conditions. In all cases structural bonds were designed to be on the rear face where a maximum temperature rise of 139°C above ambient is the maximum permissible within the fire codes. This implies the need to resist environmental loads and self weight for a period of up to two hours after fire and blast conditions commence as joint temperatures rise to 160°C. Table 4.4 shows the sensitivity to both loading and temperature in creep tests on thick adherent lap-shear specimens. At 155°C, the two part

Table 4.4. Thermal creep performance of Araldite 2004.

Material Combination	Shear Stress [MPa]	Temperature [°C]	Time to Failure [Hr]
steel/steel	1.07	200	2640
steel/steel	1.34	200	16
steel/steel	1.07	250	0
steel/GRP	0.9	100	>2000
steel/GRP	0.5	150	3
steel/GRP	0.5	155	1.5
steel/GRP	0.25	200	330
steel/GRP	0.25	203	20

epoxy adhesive used (Araldite 2004) can only sustain 3-4% (cf. table 4.3) of its average failure stress at room temperature for periods of an hour or so before failure despite the fact that it has a Tg of 90°C (high for a two part epoxy). Notice the rapid deterioration of creep performance in the 100-200°C range. The higher Tg of 120°C, associated with most single part epoxies, greatly improves high temperature performance if a hot cure is practical as long as differential adherent expansion is not a significant factor.

Where high temperatures are expected in operation, it will always be necessary to design bonded joints to be fail-safe (generally compressive) or supported by additional bolts or rivets. The development of inorganic silicate cements may provide some extension of high temperature performance in future.

4.3.4 Durability in a Marine Environment

Durability is not only a key parameter of adhesive selection but is essentially determined by the reaction chemistry of the adhesive, the adherents and the environment. Durability must be considered in terms of both the cohesive strength of the adhesive and adhesive strength at the interface and can be assessed from either accelerated ageing tests in controlled environments or long-term exposure testing. In this respect Lees [7] demonstrates, from accelerated ageing tests on a two part epoxy bond between SMC adherents, that although a well prepared (abraded and degreased) surface initially produces about 38% greater shear strength than an equivalent 'as received' surface, after 4000 hours of ageing (at 40°C in 95% relative humidity) this difference disappears with both preparations giving shear strengths of about 62% of the unaged 'as received' condition. As both ageing effects tend to level out at this point, this stress level may be taken to represent the long-term life of such a joint whichever preparation has been used. Whether this observation is more generally applicable remains the subject of further research.

In general though, long term adhesive strength is a function of the preparation of the adherent surfaces and the application of suitable primer/coupling agents, such as silanes, designed to inhibit the weakening effects of moisture at the interface. This is particularly important where metallic adherents are involved although with FRP materials long-term performance may be more related to those of the adherent resins.

In terms of long-term cohesive performance, Lees [7] states that "epoxy adhesives in their heat cured, single part form are extremely durable" although the potential of two part epoxies may be somewhat less. However, he concludes that "it is probably true to say that the durability of the substance of an epoxy adhesive - not its interface with the bonded substrate - is independent of the nature of the underlying substrate".

Acrylic adhesive, however, is far too brittle for structural purposes without being toughened with a variety of rubbery additives. If based on polyurethane chemistry these will be very susceptible to softening as the result of moisture attack and solvent absorption. In contrast those toughened with chlorosulphonated polyethylene (Hypalon) are exceedingly durable, being able to outperform the polyester composites that they were used to assemble in marine applications (AMTE(M) TM 79307). Development of these has resulted in the VOX series which are also able to bond glass directly to composite without the need for a separate primer.

The polyurethanes are sensitive to chemical attack by water, especially when warm and in the presence of metal or metal oxides, which appear to act as catalysts. For this reason polyurethanes should not generally be considered for marine structural applications in conjunction with metallic components.

4.3.5 Selection Criteria

As discussed throughout this section, the strength and performance of an adhesive will depend on details of the proposed joint configuration, the mechanical, thermal, creep and durability properties of the adhesive and adherents as well as the practicability of preparation techniques, application and cure cycles within the production environment. In the early stages of design, selection from a generic short list will probably be based on published information, supported by adhesive manufacturers, verified independently by suitably chosen small scale mechanical tests. As the design progresses, it may be necessary to conduct larger scale tests based on production prototypes for a limited number of preferred candidates. In cases where substantial structural loading is involved it will probably be necessary to consider supporting finite element stress analysis to refine the dimensions and detailed shape of the highly loaded areas of the joints to minimise highly stressed regions through increased joint ductility.

4.4 BONDING

In general, manufacturers are only too pleased to be consulted and advise on the production procedures most suitable for their particular product, although they will frequently refer customers to specialist manufacturers of large scale mixing and dispensing equipment. The observations in this section are largely drawn from small scale experiments and practical experience reported in references [7,8] and [9]. In general, care and attention to detail rather than any particular skill will pay dividends during the preparation and bonding phases. This should also extend to recognition of the chemical hazards that may exist with some of the adhesives or their individual components and the simple precautions of good ventilation (in some cases) and avoidance of skin or eye contact.

4.4.1 Surface Preparation

If it were not for accidental or intentional contamination of the surface (through release agents), the thermodynamics of epoxy and polyester composites favour the bonding process and would normally ensure a durable joint. Some form of solvent cleaning (methyl alcohol [1], styrene or acetone [10]) will therefore always be required, usually following light abrasion of the surface. However, it is suggested that in many applications a wipe with specially formulated, chemically active primers may be sufficient to neutralise the influence of mould release agents [7]. Production quality control can also be greatly assisted through the application of self indicating primer/coupling agents. These are formulated to expose contaminated regions and assist their elimination, where found, through additional light abrasion, wiping dry and retreatment.

In thick, hand-laid FRP, the uneven rear face makes light abrasion unreliable. The alternative use of shot blasting has been found to damage the surface layers of reinforcement as well as embed unwanted debris in the surface. Instead, the use of a nylon peel ply in the final surface lay-up has been shown to improve joint shear strength by up to 20% as long as surface degreasing is used after the peel ply removal [8]. Some benefit is also thought to come from the thinner resin layer that results from peel ply application.

4.4.2 Dispensing, Clamping and Curing Procedures

In general, for large scale applications reliance on hand mixing of two part adhesives is likely to be both impractical and difficult to control. The result will not only entrap air to form voids but often suffer from substantial variation in component proportions leading to areas of incomplete cure. Laboratory experiments have indicated that a 15% increase in shear strength may result from automatic mixing [9]. Practical dispensing equipment ranges from 200-400 ml twin cartridge, hand-held (manual or air driven) dispenser guns to large 50-100 litre fully automated dispenser systems, all using disposable, opposed helix nozzles for mixing at the point of application. The pseudo-one part acrylic adhesives may offer some advantages over the normal two part varieties in offering the option of spreading one component on each joint component so that mixing will occur only when the parts are brought together.

In all cases, some form of jigging will be required although this may be minimised through the use of self-jigging joints based on extrusions or pultrusions illustrated in figure 4.4. De-jig or handling times vary greatly depending on adhesive type and temperature and a useful guide to cure rates is offered by Lees [7]. In general, where large components are being assembled, a long cure time (measured in hours) is beneficial. In such cases it is often more practical to arrange for overnight cure before de-jigging to ensure no damage to the joints result from handling.

The presence of some heat during cure will not only reduce cure time but will improve adherent wetting and increase joint strength through more extensive cross-linking. It will not generally be practical or economic to provide this for large FRP structures although use can be made of subsequent oven treatments for smaller components if structural integrity is ensured through jigs, bolts or rivets in the meantime.

4.4.3 Quality Assurance

As is the case with FRP materials in general, there are few straightforward techniques available for the non-destructive testing of bonded joints. Careful attention to procedure, verified by type testing, remains the normal approach at present, supplemented by careful visual examination of exposed spew fillets, which are the highly stressed areas of most joints. In translucent FRP materials, the use of black adhesives allows the presence of significant debonded ares of a joint to be clearly seen through the thickness of the adherent. These may also be detected by tap tests or ultrasonics, but the detection of small voids is neither generally practical nor of known significance.

4.5 OPPORTUNITIES FOR APPLICATION OF ADHESIVES

As FRP components, particularly those produced to exacting dimensions through pultrusion or other automated processes, become more readily available, the opportunity to use adhesives either on their own, or in combination with other joining techniques, become endless. At the time of writing, a 120 m (63 m central span) cable stayed footbridge is nearing completion over the River Tay on the Aberfeldy golf course [11]. The entire deck is formed from interlocking pultruded sandwich panels (600 x 80 mm section) which are bonded together on site using a two part epoxy adhesive by teams of 'unskilled' university students. The patented panels can just as well be foam insulated and assembled to form bonded structural walls.

Smith [10] offers a wide range of possible applications in the topsides of future warships where cheap, lightweight steel framed hybrid GRP structures have been shown to offer considerable design potential. This approach, either all FRP or hybrid, is equally applicable to offshore topsides where the beneficial effects of lightweight GRP for blast and fire-resistance have been applied recently in both new and refurbished structures following the Piper Alpha disaster [9]. Opportunities also exist in the design of FRP submersibles where stiffener/shell attachments are permanently in compression - taking maximum benefit from the strength of adhesives.

Of particular interest are more complex design solutions such as the dual adhesives within the traditional hand-laid, top hat stiffener to shell attachment used in British GRP MCMVs. Smith [10] describes the clear benefits to be obtained from including a substantial bead of resilient acrylic adhesive at the root of flange attachment which otherwise relies on the base polyester resin for its attachment. The use of this design eliminates the need for bolted reinforcement of this joint to resist shock loading.

4.6 REFERENCES

1] Shields, J., "Adhesive Bonding", *Engineering Design Guide*, **02**, OUP, Oxford, 1974.
2] Lees, W.A., "Adhesives in Engineering Design", The Design Council, London, 1984.
3] Shields, J., "Adhesives Handbook", 3rd Edition, Butterworth, London, 1984.
4] Skeist, J., (ed.), "Handbook of Adhesives", 3rd Edition, Van Nostrand Reinhold, New York, 1990.
5] Brewis, D.M., Briggs, D., (ed.), "Industrial Adhesion Problems", Orbital Press, 1985.
6] Lees, W.A., "Designing for Adhesives", Proc. Conf. *Materials and Engineering Design*, London, May 1988.
7] Lees, W.A., "Recent Developments in Composite Bonding with Particular Reference to Large Structures and Unprepared Surfaces, Proc. Intl. Conf. *Advances in Joining Plastics and Composites*, TWI, Bradford, June 1991.
8] Cowling, M.J., Hashim, S.A., Smith, E.M., Winkle, I.E., "Adhesive Bonding for Marine Structural Applications", Proc. Intl. Conf. *Polymers in the Marine Environment*, IMarE, London, October 1991.
9] Winkle, I.E., Hashim, S.A., Cowling, M.J., "Lightweight, Fire Resistant, GRP/Steel Composites for Topsides", Proc. Intl. Symp. *Marine Structures*, Shanghai, September 1991.
10] Smith, C.S., "Design of Marine Structures in Composite Materials", Elsevier Applied Science, London, 1990.
11] Robbins, J, "Links to a Tee", New Civ. Engr., 13th August 1992

5 PRACTICAL DESIGN OF JOINTS AND ATTACHMENTS

5.1 BACKGROUND

5.1.1 Need for Joints

Joints become necessary in a structure for three main reasons. These relate to production/processing restrictions, the need to gain access within the structure during its working life, and repair of the original structure.

Typical production/processing restrictions arise because of the need for:

i. Large and complex structures which cannot be formed in one process thereby needing several components to be joined to produce the completed structure. Considerations that limit process size include exotherm, resin working time, cloth size and "drapability", and mould accessibility and release limitations.

ii. Splitting the load path (and hence the fibre path) around the structure. This typically involves the addition of stiffeners and bulkheads. Generally these out-of-plane elements cannot be formed at the same time as the rest of the structure and so need to be joined to it.

Considering access and repair considerations, if components within the structure require regular servicing then the structural elements that obstruct access need to be joined to the remaining structure in such a way as to allow them to be removed with reasonable ease. If the hidden components require only very occasional treatment (such as removal after a major breakdown) then the structure can be cut out as necessary and treated as a repair. Here the jointing method can be considered to be permanent.

5.1.2 Requirements

To fulfil its role a joint must meet one major requirement; the integrity of the overall structure must not be impaired by the presence of the joint. "Structural integrity" can be defined in several ways, depending on the particular application, and can include one or more of the following:

a) Strength - in tensile or compressive, shear or through-thickness directions. A joint must be at least as strong as the surrounding structure.
b) Stiffness or flexibility - if there is a differential in stiffness between a joint and the surrounding structure then stress concentrations will occur in the joint, the surrounding structure, or both, depending on the particular geometry and loading.
c) Water- (or air-) tightness - if the purpose of the structure is to retain (or prevent the ingress of) a fluid of one sort of another then obviously any joints on the surface of the structure need to be equally secure.

Another requirement which needs to be considered at the design stage is the economics of producing the joint. In large and complex structures the joints can constitute a significant proportion of structure weight and their manufacture can be expensive. It is necessary to ensure that material and labour requirements are kept to a minimum and the technology used to produce the joint is compatible with that used elsewhere in the structure. Care should be taken to balance the economic advantages of reduced weight against the possible additional costs incurred in producing a high performance joint.

5.1.3 Types of Joints

The range of joints used in both single skin and sandwich structures can be split into two basic types, namely in-plane and out-of-plane joints.

A. In-plane or Butt Joints:
These are used whenever two plate elements need to be joined. As shown in figure 5.1, these can be either scarfed joints or lapped joints. Scarfed joints are adhesively bonded. They can be symmetric or asymmetric and can be used to bond sandwich panels providing the skins are sufficiently thick. Stepped-lap-joints may also be considered in this category. Lapped joints

may be bolted, adhesively bonded, or both. They can be single- or double-lap, and can be used to bond sandwich panels, especially when the skins are thin.

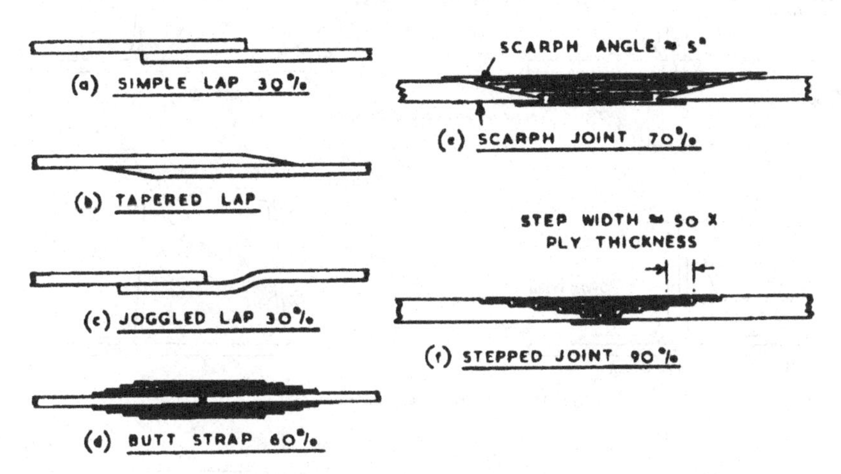

Figure 5.1. Typical arrangements and efficiencies of in-plane joints.

B. Out-of-plane Joints:
Typical examples are shown in figure 5.2 and include frame-to-shell connections and bulkhead-to-shell connections. The former mainly employs top hat stiffeners which are usually moulded in situ onto the cured shell. It is also concerned with extruded/pultruded section stiffeners which are preformed and bonded in place using adhesives and/or boundary angles. Bulkhead-to-shell connections differ from the top hat stiffener frame-to-shell connection in that both elements are formed before the joint is fabricated, and access is available from both sides of the joint.

5.1.4 Bonded vs Bolted Connections

The choice depends very much on the application being considered. A bonded connection provides a greater surface area to transmit load. This ensures that all fibres at the joint interface are used to carry load so that stress concentrations are reduced. They are cheaper and easier to produce and can be formed from one side of the panel. However, some environmental control is usually required during the construction process. Another shortcoming is that when initial failure occurs in a purely bonded joint it can propagate easily since there are no fibres

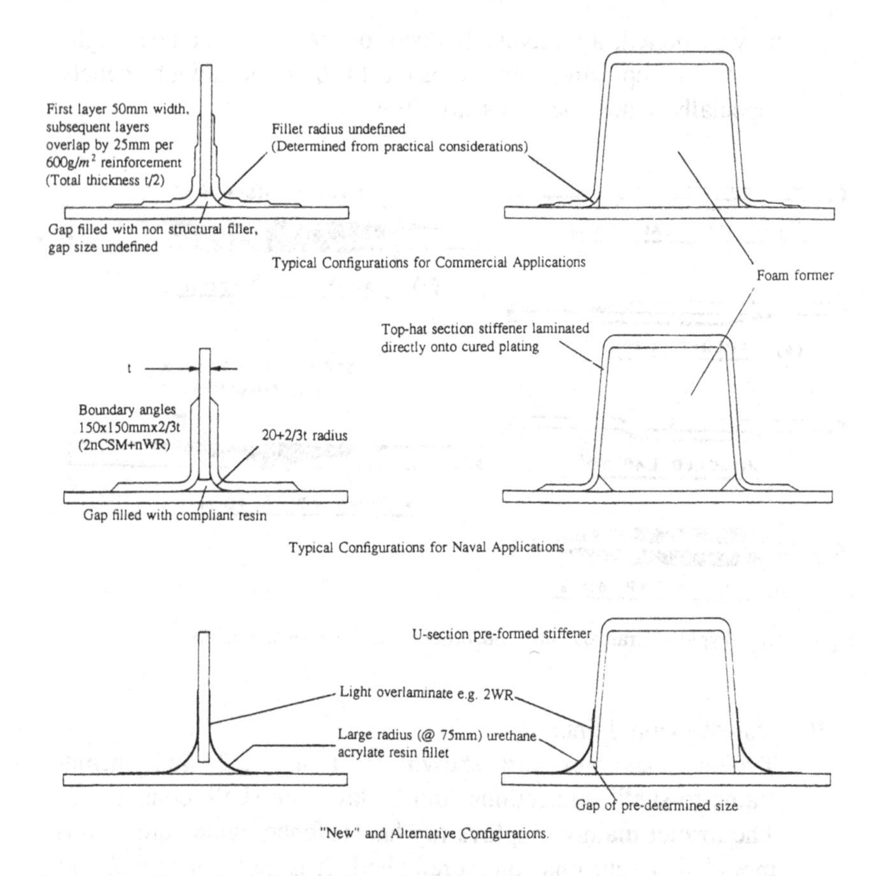

Figure 5.2. Typical configurations of bulkhead-to-shell joints and top hat stiffeners.

across the joint to act as crack arrestors. Such joints are permanent and cannot easily be removed.

Bolted connections provide a strong link across the joint interface; they are easily removed, and can usually be formed under adverse environmental conditions. When used in conjunction with an adhesive, the bolts can act as crack arrestors in the event of initial failure. However, since the load is transmitted through a small area, stress concentrations occur that can lead to early initial failure. They require access from both sides of the joint, are heavy and can be expensive to produce.

The performance of both bonded and bolted joints is sensitive to the lay-up of the composites being joined, particularly if they are formed using a large proportion of unidirectional fibres. This must be considered when using single-lap or stepped-lap-joints and, to a lesser degree, other forms. Typical marine composites contain high proportions of woven or randomly oriented fibres, which are less sensitive to the jointing method used.

For bonded joints, the geometry should be arranged so that the layer at the joint interface contains fibres that are oriented in the direction of the local loads. This can usually be arranged and thus is not an initial design consideration.

For bolted joints, the combination of layers through the thickness of the laminate should ideally be arranged so that the full bearing strength of the laminate is achieved before pull-out or pull-through. This needs to be considered at the initial design stage.

5.2 IN-PLANE OR BUTT JOINTS

5.2.1 Features and Purpose

As mentioned above, these joints are used to join structural elements together, attach removable panels to the main structure and make repairs to a structure. They are used in two basic forms, namely scarfed and lapped. The decision as to which form to use depends upon the particular application and the thickness of the adherents being joined. The relative advantages and disadvantages of different joint types are summarised in table 5.1.

Sandwich panels generally have thinner skins which can make scarfing much more awkward. However, even when the panels are in bending, the loading on the skins remains in-plane, as the panels are much stiffer than the individual skins. So the bending and stress concentration effects associated with lapped joints are much reduced.

A considerable amount of work has been carried out on the analysis of butt joints in the past, both experimentally and theoretically, for the case of in-plane loading. This work has been thoroughly reviewed elsewhere [2,3] so that task will not be repeated here. Only the influence of different design variables and analysis methods are briefly outlined in the following three sections.

5.2.3 Design Variables

Assuming that the magnitude and mode of loading is known, a series of design variables can be identified for any given joint configuration. Firstly, considering the structural elements being joined, the variables that need to be considered are:

a) Fibre type and form

Table 5.1. Summary on in-plane joint types.

Joint Type	Advantages	Disadvantages
Symmetric scarf	Best in flexure Good in tension Uses less material Similar flexibility to original structure	Expensive in labour Difficult on thin laminates Needs access from both sides Permanent Joint
Asymmetric scarf	Good in flexure Good in tension Needs access from on side only Similar flexibility to original structure	Expensive in labour Difficult on thin laminates Uses more material Permanent Joint
Double-lap	OK in flexure Good in tension Easy on thin laminates Cheaper in labour Removable joint if bolted	Needs access from both sides Uses more material Stiffer than original structure
Single-lap	Needs access from one side only Cheapest in labour Easy on thin laminates Removable joint if bolted	Poor in flexure Poor in tension Uses more material Stiffer than original structure

b) Resin type
c) Fibre orientation/stacking sequence
d) Laminate material properties (moduli, UTS, shear strength, etc)
e) Laminate thickness

Next, considering the type of connection used, for bolted connections design variables include (figure 5.3):

a) Bolt diameter
b) Hole size and tolerance
c) Clamping force
d) Washer size
e) Pitch of holes
f) End distance
g) Side distance
h) Back pitch if several rows are used
i) Joint type (single or double overlap)

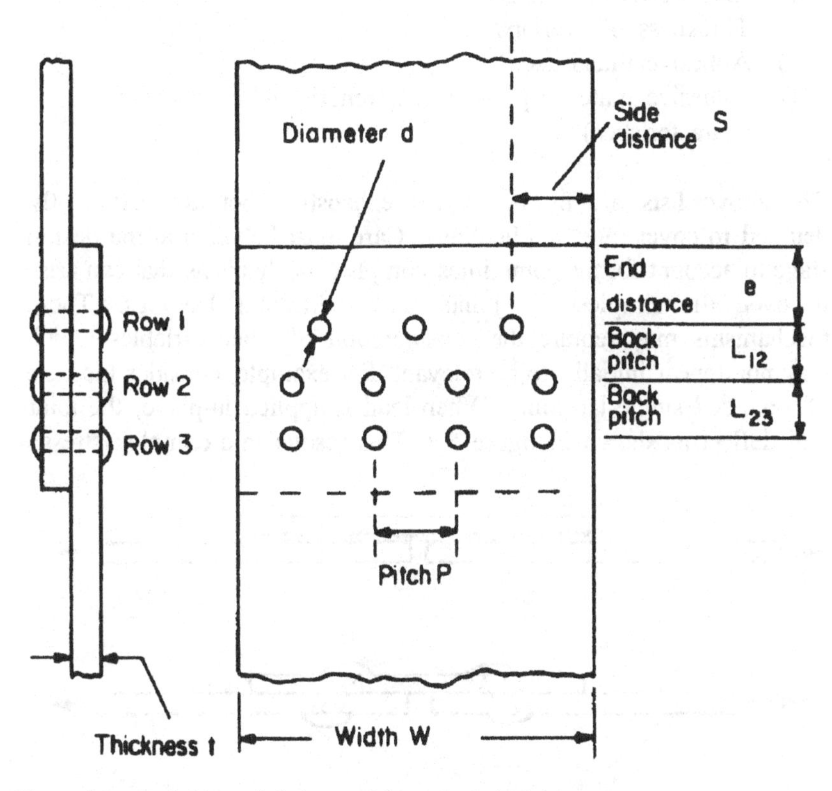

Figure 5.3. Definition of design variables for bolted joints.

For bonded connections two further subsets of variables exist, namely those that apply to scarfed or stepped-lap-joints and those that apply to lapped joints. The design variables which apply to scarfed or stepped-lap-joints include:

a) Angle of scarf or length of cutback
b) Depth of cutback (influenced by the stacking sequence of the

structural elements being joined, as mentioned in section 5.1.4 above)

c) Symmetric or asymmetric joint
d) Lay-up of the joining composite

Those which apply to lapped joints include:

a) Single- or double-lap
b) Length of overlap
c) End profile of overlaps
d) Thickness of overlaps
e) Adhesive thickness
f) Adhesive material properties (strength and stress-strain characteristics)

The above lists are by no means exhaustive but are sufficiently detailed to cover most applications. Care must be taken at the design stage to account for the sometimes complex mechanisms that can arise in even the simplest of joints under different loadings. These mechanisms may require the consideration of other variables which may not appear initially to be relevant. For example, consider the case of a bonded single-lap-joint. When load is applied in-plane, the joint will deflect as shown in figure 5.4. This results in a complex stress

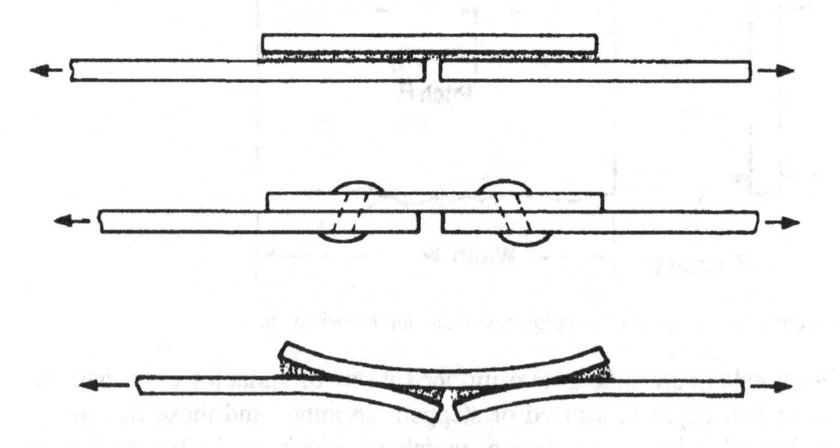

Figure 5.4. Deflection characteristics of a single-lap joint.

state that includes through-thickness tensile stresses in the adhesive and the adherents, shear in both the longitudinal and through-thickness directions, and longitudinal stress in the adherents. The stresses in the adhesive vary through the thickness. This reinforces the need to have well identified analytical and modelling techniques which are able to cater for such mechanisms.

Clearly, if a modelling technique is to account for all the above mechanisms and design variables it will be complex and cumbersome, and thus unsuitable for use in a design context. Many of the out-of-plane material properties that are required for these techniques are unlikely to be available for a given laminate, necessitating an extensive materials testing programme prior to design. This is impractical in most circumstances. Hence it is necessary to eliminate many of the variables when considering a particular problem. In previous work, the behaviour of a few basic joints in a limited range of materials has been analysed and the influence of the more important variables has been determined. These results can be used in a qualitative sense to predict the behaviour of other joints.

5.2.4 Modelling Techniques for Bolted Joints

Consider the case of a bolted single-lap-joint loaded in-plane. A simple modelling technique is to take the bearing stress of the composite (σ_b), its thickness (t), and the diameter (d) and number of bolts (n) used;

$$\text{Load (P)} = \sigma_b \ t \ d \ n. \qquad \text{For a unit width Capability of joint.} \qquad (5.1)$$

Generally, the load capability and thickness of the composite being joined will have already been determined from other structural considerations, at least on a preliminary basis. If the criterion is to reduce the number of bolts used, it is necessary to achieve as high a bearing stress and as large a diameter as possible. These variables are, however, linked, both with each other and with other design variables. Figure 5.5 shows that bearing stress reduces as d/t increases if no lateral restraint is applied to the joint, and remains constant if a lateral restraint is applied. It is further suggested that d/t should not exceed 1 if full bearing strength is to be achieved. When considering small diameter bolts the shear strength of the fastener must be considered. Reference [6] provides data sheets on fastener performance in several applications but the simplest formula would be;

$$\text{Shear Strength of Fastener} = P \ n \ \pi \ (d/2)^2 \qquad (5.2)$$

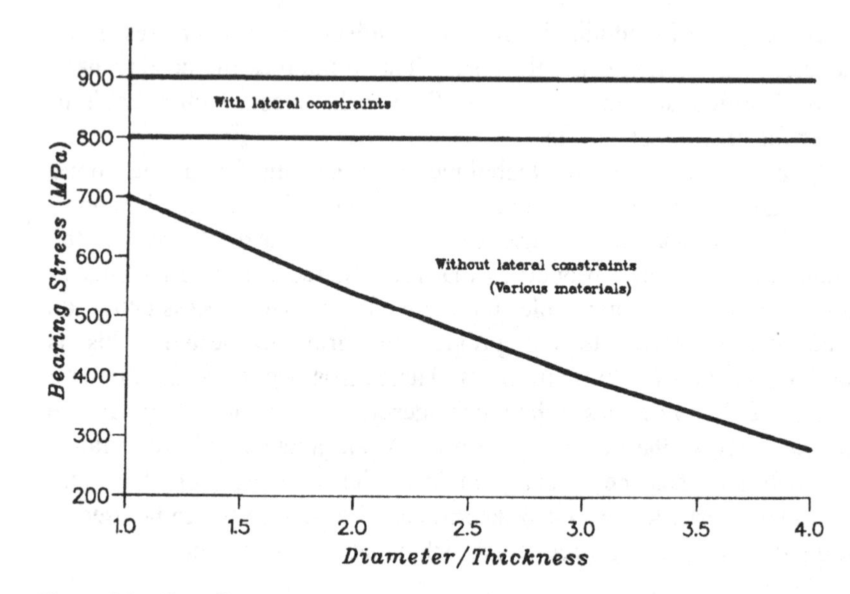

Figure 5.5. The effect of d/t ratio on bearing strength (Data from [5]).

Thus an upper and lower limit on d can be arrived at. All that remains is to ascertain the achievable bearing stress for the given joint. Unfortunately this is rarely found in material data since it is so dependent on geometrical variables. Maximum values for bolted connections have been quoted [2] as being between 800 and 930 MPa for CFRP, 550-700 MPa for unidirectional GRP, and 200 - 600 MPa for woven or CSM GRP. Generally, failure can occur in three ways by shear, bearing, or tension. Figure 5.6, for example, shows the relationship between bearing stress and lay-up in CFRP. Quasi-isotropic laminates will fail in bearing mode at the highest loads. Other factors affecting bearing strength are clamping pressure, joint width or pitch, end distance, and load direction. Curves similar to figure 5.6 can be found in reference [2] for these factors. Stacking sequence has little effect on the bearing strength of bolted joints if the plies are homogeneously mixed through the laminate. If the laminate is blocked (i.e. all layers of a given orientation are grouped together) significant reductions of up to 50% can result [7]. It is also important that the bolt is a good fit in both the hole and any washers used.

A reasonable prediction of bearing strength can be made by considering the effects of the criteria mentioned above. By completing a simple iterative process using the two equations, the joint parameters can be derived, having accounted for the laminate properties (type,

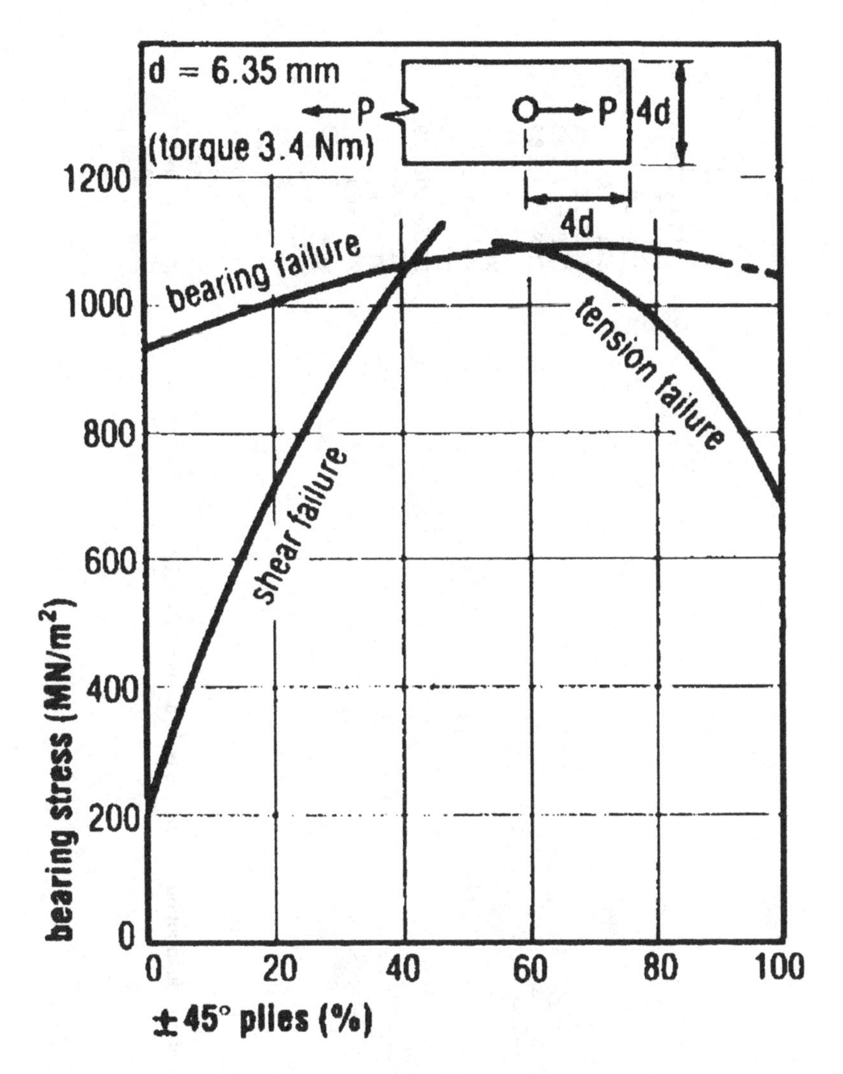

Figure 5.6. Influence of fibre-orientation on failure mode of bolted joints in 0/+-45 CFRP. (From [4]).

orientation, material properties, thickness), bolt diameter, clamping force, pitch of holes, end distance, and side distance. It should be borne in mind that the relationships illustrated above are derived experimentally from in-plane loading and, whilst they do account for out-of-plane deflections, care must be taken when extrapolating any results from one configuration to another.

The same principle can be applied to a double-lap bolted joint,

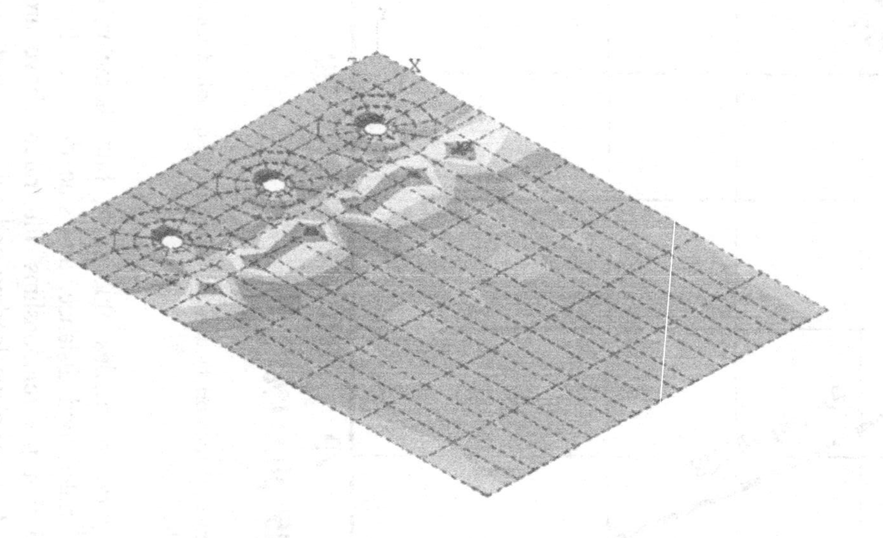

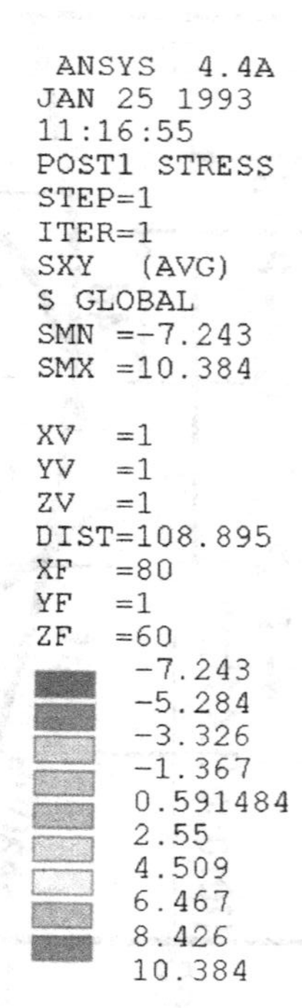

Figure 5.7. Typical stress distribution around bolts in a layer of CFRP.

since each lap is usually half the thickness of those in a single-lap-joint. The reduced out-of-plane deflections in a double-lap-joint means that failure is more likely to occur in bearing or tension and hence this is a better jointing method particularly for highly anisotropic laminates. One change to the above equations is that the shear stress in the fastener is half that found in the single-lap-joint, and hence a different optimum arrangement will be arrived at.

If greater detail is required or more complex out-of-plane loading is anticipated, then it may be necessary to carry out a numerical analysis of the joint using, for example, the finite element method. This method allows a layer by layer analysis of the stress distribution within a joint. Many commercial packages are available and, with modern graphical interfaces, are as quick and easy to use as the classical analytical methods without being constrained by the assumptions these methods make. Figure 5.7 shows a typical stress contour plot of a bolted single-lap-joint being subjected to flexure. Extensive experimental and numerical analysis in CFRP [8] has shown the performance of the joint to be sensitive to stacking sequence, fastener diameter, pitch, and clamping pressure. These variations are non-linear in nature and it is recommended that each problem of this type is analysed independently.

5.2.5 Modelling of Adhesively Bonded Joints

Several classical analytical methods exist for analysing bonded joints between isotropic plates loaded in tension [3]. When applying these methods to composites care must taken to account for the anisotropic nature of the adherents and the variation in modulus through the thickness. Consideration should also be given to thermal loadings resulting from the curing cycles. Generally, the simpler the analytical method, the less accurate it will be when applied to single-lap-joints due to the out-of-plane displacements described in section 5.2.3 above. However, for the scarfed, stepped-lap, and double-lap joints where these out-of-plane displacements are less significant, reasonable correlation between analytical results and experiments can be achieved [9-11] providing any non-linear behaviour of the adhesive is accounted for.

For practical design purposes, however, even the simplest shear lag theory is impractically complex and of limited application since the assumed boundary conditions are not modelled in real life. Of much greater use to the designer are the guidelines produced by various classification societies and other sources [12-16], and the results of parametric studies that have been completed. Significant features are outlined below:

a) The stresses in the adhesive reduce and become more uniform as the adhesive modulus is reduced, the adherent stiffness is increased and the overlap length is increased.
b) Tapering the ends of the adherents on double-lap-joints and including an adhesive spew fillet on single-lap-oints reduces the maximum shear and direct stresses in the adhesive.
c) The adherents at the joint interface should be identical if possible. If this is not possible then the in-plane and bending stiffnesses should be matched.
d) If a single-lap-joint is restrained to prevent rotation, and hence the formation of peel stresses (e.g. on sandwich skins), its strength is improved and thus it can be applied to thicker adherents. Similar improvements can be made if a through-thickness clamping force is applied at the ends of the joint (see figure 5.8).

FM-400 bonded joints (width = 1.00 inch)	failure mode	joint eff
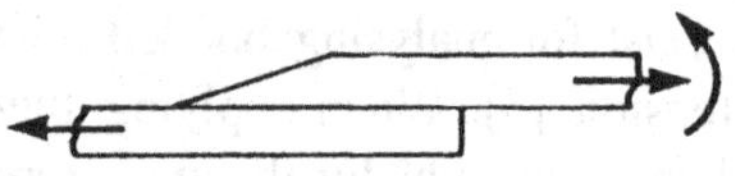 control	peel and shear failure	0.52
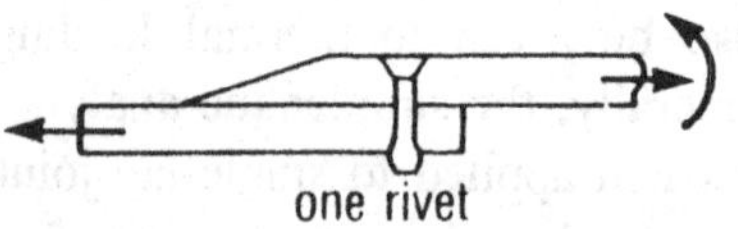 one rivet	shear and patch net tension failure	0.73
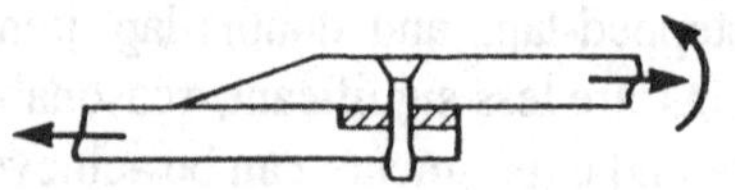 undercut and one rivet	shear and rivet head pull-through failure	0.78

Figure 5.8. The effect of detailed design in reducing peel stresses.

e) Plies at the joint interface should be aligned in the direction of the applied load where possible.
f) The length of overlap in a double-lap or stepped-lap joint must be long enough to allow the "elastic trough" to develop (see

figure 5.9). This is the name given to the region where the shear stress in the adhesive is below its elastic limit. Any increase above this results in little increase in strength.

g) The adherents in a scarf joint should ideally have knife edges. Finite tip thicknesses cause strength reductions of up to 25%.

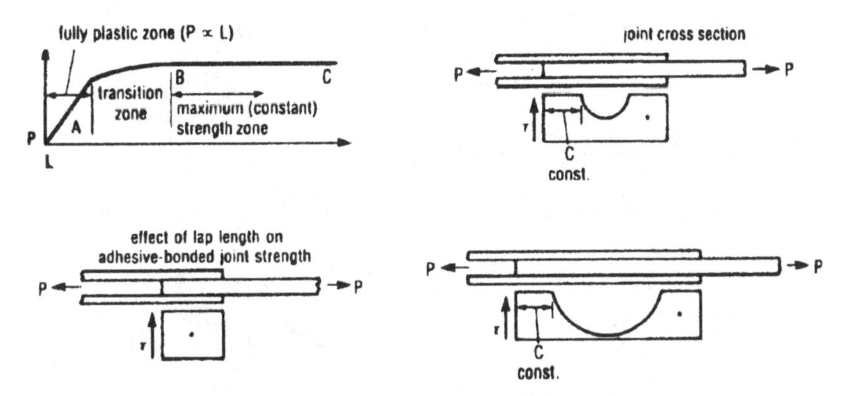

Figure 5.9. Effect on lap length on strength and adhesive shear stress distribution of bonded double-lap joints.

The finite element method has been used extensively to analyse bonded joints. This allows analyses to be completed without the need to make the simplifying assumptions necessary for the classical analytical methods, and can be used to tackle problems that cannot be solved using classical analyses. An example of a simple plain strain analysis of a double-lap-joint is shown in figure 5.10. This illustrates the high peeling and shear stresses at the end of the lap.

5.3 OUT-OF-PLANE JOINTS I: FRAME-TO-SHELL CONNECTIONS

5.3.1 Types of Connections

As mentioned in section 5.1.3 above, the principal type of stiffener used in marine composite structures is of a top hat configuration. It is usually formed in situ over a former onto the cured panel or shell. This imposes some limitations on the design of the joint between the stiffener and the panel (see section 5.3.2 below). Nevertheless several different designs have been developed, principally in an attempt to prevent stiffener debonding under extreme loads. One recent development being considered [17] is that of preforming top hat sections and bonding these cured sections onto the shell. The major

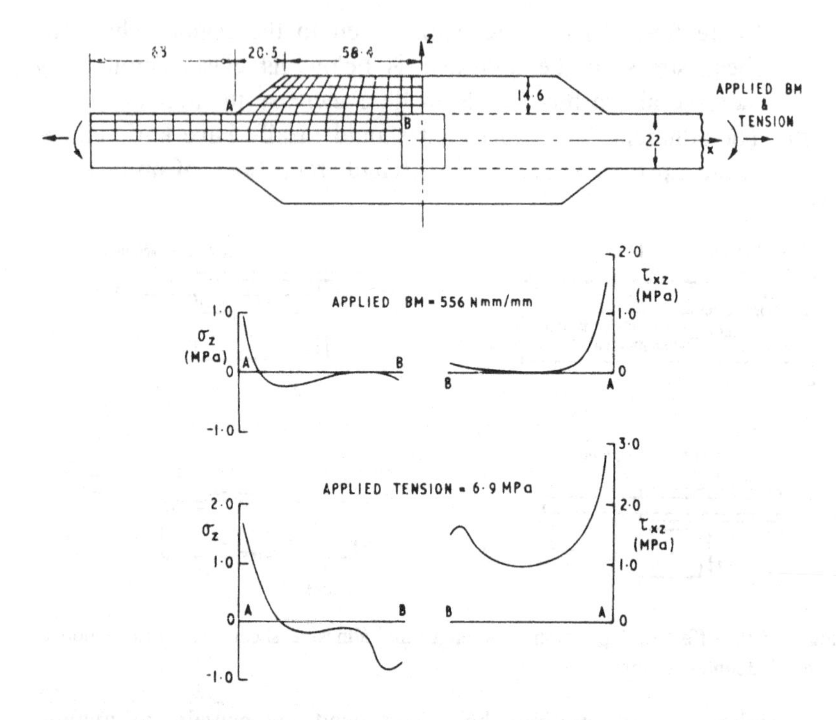

Figure 5.10. Stress analysis of a double-lap joint.

benefits of this development are much reduced production costs and greater flexibility in the design of the joint.

5.3.2 Design Variables

Taking a typical example of a top hat joint, given in figure 5.2, the variables available to the designer are:

a) Radius of fillet
b) Backfill angle of fillet
c) Fillet material
d) Thickness of overlaminate
e) Length of overlap of overlaminate
f) Lay-up of overlaminate
g) Overlaminate resin.
h) Number, position, and make-up of additional plies
i) Number, position, and size of reinforcing bolts

However, the current practice of forming the stiffeners in place means that the variables associated with the overlaminate (namely items d, f and g above) are fixed since the overlaminate around the joint is just an extension of the web of the stiffener, whose dimensions are governed by global stiffener requirements. The development of preforming the top hat section removes these constraints since the overlaminate is separate from the structural elements being joined. For economic reasons additional plies and reinforcing bolts (items h and i) should be avoided if possible.

If the stiffener is preformed and bonded in place these variables remain flexible, and the gap between the stiffener and the shell becomes an additional variable. However, the ability to control the backfill angle of the fillet is lost.

5.3.3 Modelling Techniques

As far can be ascertained no analytical methods exist to model the complex behaviour of a joint of this nature. Empirically based design guidelines are available [12-16] but these consider only a limited number of variables.

A considerable amount of experimental work has been conducted, especially for naval vessels, to determine the effects of some of these design variables [17-20]. When considering this work it should be borne in mind that the driving design criterion for these vessels is that they should be able to withstand high levels of explosive loading. This loading applies high through-thickness tensile forces to the joint which will not be seen in other applications. Initial work showed that an all polyester construction was prone to early failure in this mode. By reinforcing the joint with bolts the ultimate performance is much improved but the bolts do not prevent initial failure.

Further work investigated the use of stitched cloth and the use of compliant resins at the joint interface and for the fillet material. It was found that increasing the compliance of the joint region markedly improved the resistance of the joint to initial failure. This theme has been pursued further leading to the removal of the expensive bolted reinforcement.

A stress analysis of a frame-to-shell connection was made using 2-D finite elements [21]. The length of overlap was varied as well as the addition of internal and external flanges. Several different loadings arrangements were applied. The conclusions drawn were that the magnitude and distribution of the tensile debonding stresses were very sensitive to the form of the applied load. For loading arrangements that caused high tensile stresses in the root of the joint (the point of initial failure observed in experiments) only the internal flange

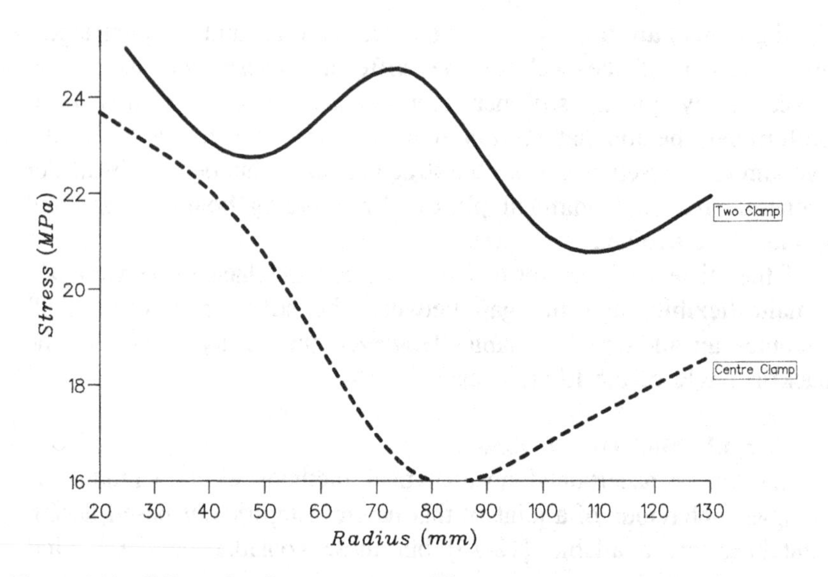

Figure 5.11. Effect of radius on stress in fillet.

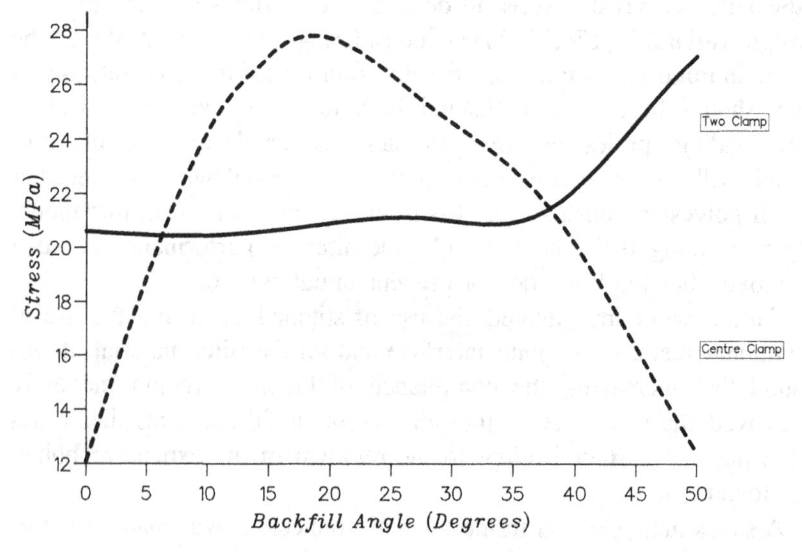

Figure 5.12. Effect of backfill angle on stress in fillet.

significantly reduced these stresses, i.e. the length of the flange overlap is not a significant design variable.

A parametric study has been completed [17] to assess the feasibility of the preformed top hat design and to determine the effect of different design parameters. Figures 5.11 to 5.15 show the variations of internal stresses relating to each variable. These results can be used to optimise the joint parameters for the particular load level anticipated.

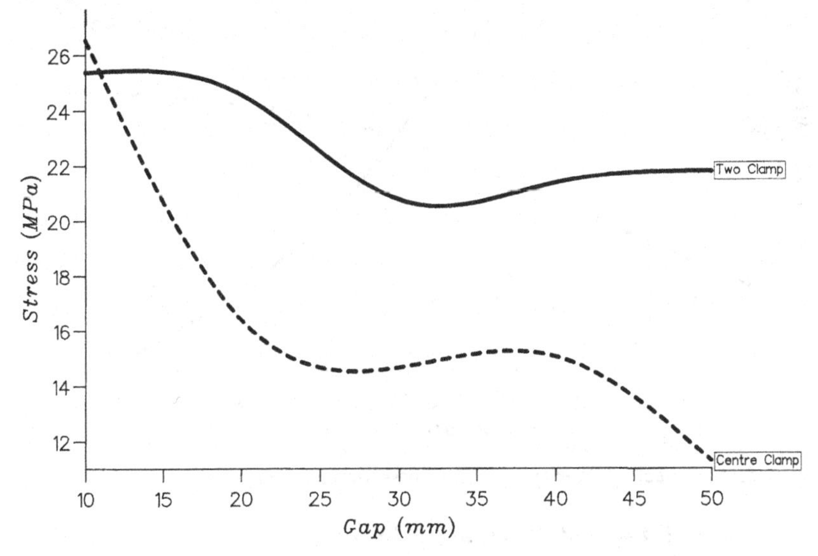

Figure 5.13. Effect of gap on stress in fillet.

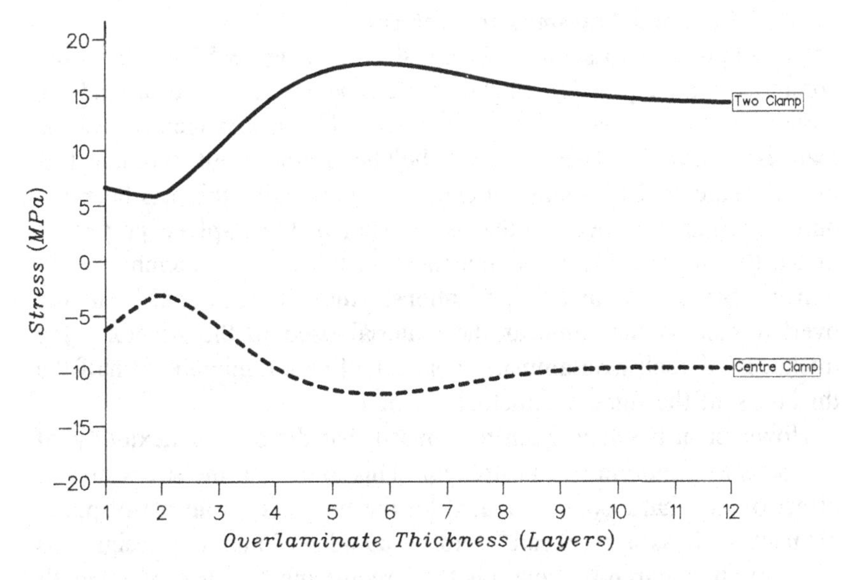

Figure 5.14. Effect of overlaminate thickness on through-thickness stress in overlaminate.

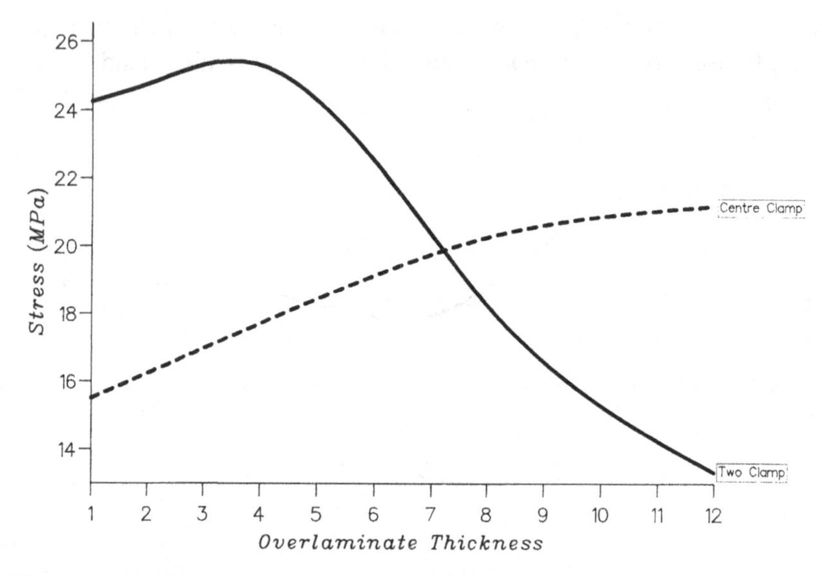

Figure 5.15. Effect of overlaminate thickness on stress in fillet.

5.4 OUT-OF-PLANE JOINTS II: BULKHEAD-TO-SHELL CONNECTIONS

5.4.1 Load Transmission Features

Typical bulkhead to shell joints are shown in figure 5.2. One feature common to all types, is that the overlaminate is wrapped around the corner linking the two structural elements. This is arranged so that the load is transmitted between the bulkhead and shell through this overlaminate in its in-plane direction. Traditionally this has been the only criterion considered - that is the sum of the in-plane properties of the two overlaminates is equivalent to the weakest member being joined. Since, in most applications, the material used in the overlaminate is the same as the material used in the structure, the thickness of each overlaminate is specified as a minimum of half the thickness of the thinner structural element.

However, it has long been recognised that the overall flexibility of a joint is as important as its strength. This is valid if the stress raising effect of the "hard point", created by the presence of an out-of-plane element such as a bulkhead, is to be avoided. Thus, the designer is faced with apparently contradictory requirements, that of strength (requiring a thick overlaminate) and flexibility (requiring a thin overlaminate).

5.4.2 Failure Modes

This contradiction is ably illustrated by considering the failure modes associated with these joints. The two elements carrying load are the overlaminate and the fillet. For both elements to be exploited efficiently they should both reach their respective failure stresses (both in-plane and through-thickness in the case of the overlaminate) at the same load. If the overlaminate is relatively thick (as in most cases) then it is also stiff. As load is applied to the joint, high through-thickness stresses develop in the corner of the joint and it fails by delamination. As the load is increased, further layers will delaminate until the overlaminate delaminates from the fillet. As soon as this occurs, the unsupported fillet fails. During this process the in-plane stress in the overlaminate is much lower than its failure value. Similarly, prior to failure, the stress in the fillet is also lower than the failure value of the material. This failure process has been observed by Hawkins et al. [22].

Clearly, to fully exploit the materials used, it is necessary to reduce the through-thickness stress, whilst allowing the in-plane stress and the fillet stress to increase. Traditionally, attempts have been made to reduce the delamination failure by increasing the overlaminate thickness. This is done to increase the stiffness of the joint, reduce deflections, and hence reduce the through-thickness stress. Also different laminating resins have been used with improved through-thickness properties. However, this approach reduces still further the in-plane stress and the fillet stress, so is less efficient in its use of materials, and creates a worse "hard point" in the structure. Any design method must be able to analyse the joint to quantify these stresses, and be able to determine the effect of design variables on them.

5.4.3 Design Variables

The design variables that relate to this form of joint are similar to those for top hat stiffeners above:

a) Radius of fillet
b) Fillet material
c) Thickness of overlaminate
d) Length of overlap of overlaminate
e) Lay-up of overlaminate
f) Overlaminate resin
g) Gap between bulkhead and shell

h) Edge preparation of bulkhead
i) Number, position, and make-up of additional plies
j) Number, position, and size of reinforcing bolts

Here, again, it is desirable to avoid using additional plies or reinforcing bolts if possible.

5.4.4 Design Modelling

Numerical analysis using, for example, the finite element method, is the easiest way to assess these relatively complex problems and account for all the design variables. A series of studies have been done [23] to investigate the impact of these variables, and which resulted in the new type of joint design shown in figure 5.2. This design exploits the low modulus and high strain to failure of the fillet material which allows large displacements and rotations to accommodated without failure.

The mechanism of this joint is as follows. As load is applied, the thin overlaminate bends so that the load can be transferred in its in-plane direction, as before. However this time, because the laminate is thin, and therefore flexible, the through-thickness stresses remain low. The out-of-plane deflections are accommodated by the flexible resin. The large radius gives the overlaminate a greater leverage on the joint so the in-plane stresses remain reasonable despite the overlaminate being thin. The actual values of overlaminate thickness and fillet radius can be optimised so that the ideal of ultimate failure occurring in the overlaminate (both in-plane and through-thickness) and in the fillet at the same load can be achieved. This effectively removes the contradiction that develops when using the traditional approach.

These two variables (i.e. overlaminate thickness and fillet radius) are by far the most important and their effects are summarised in table 5.2. Other variables have also been considered but were found to be less significant. Generally, variables should be adjusted to make the joint as flexible as possible whilst keeping stresses low. This can only properly be monitored by using the finite element method or by extensive, and expensive, experimental work.

The method just described can be equally well applied to the joining of sandwich panels but several other aspects must be considered. Principal amongst these is the need to keep the joint as flexible as possible to reduce through-thickness stresses between the skin and the core of the sandwich panels, which would otherwise

Table 5.2. Impact of geometry and material variations.

Response Feature	In-Plane Stress in Laminate	Through-Thickness Stress in Laminate	Principal Stress in Fillet
Increasing Thickness of Overlaminate	Decreases	Increases	Decreases
Increasing Radius: Overlaminate	Decreases	Decreases	Decreases
Increasing Radius: Pure Fillet	-	-	Approx. Same
Impact of overlaminating on a Fillet	-	-	Decreases

delaminate. The same numerical analysis methods can be used but with the different limitations applied a different optimum solution will develop. The core of the sandwich panels can be reinforced or removed in the region of the joint to overcome this problem - see figure 5.16 - and detailed finite element analysis can ascertain the stress patterns [24,25].

5.5 STIFFENER INTERSECTIONS

5.5.1 Purpose

In large 3-D structures that are subjected to complex, multidirectional loads it is sometimes necessary to stiffen the structure in two orthogonal directions. These stiffeners may be forced to intersect with one another. This should be avoided if at all possible. Stiffener intersections are fiddly to fabricate, expensive, heavy, and produce stress concentrations.

5.5.2 Design Features

The design features depend on the form of stiffeners being joined. In

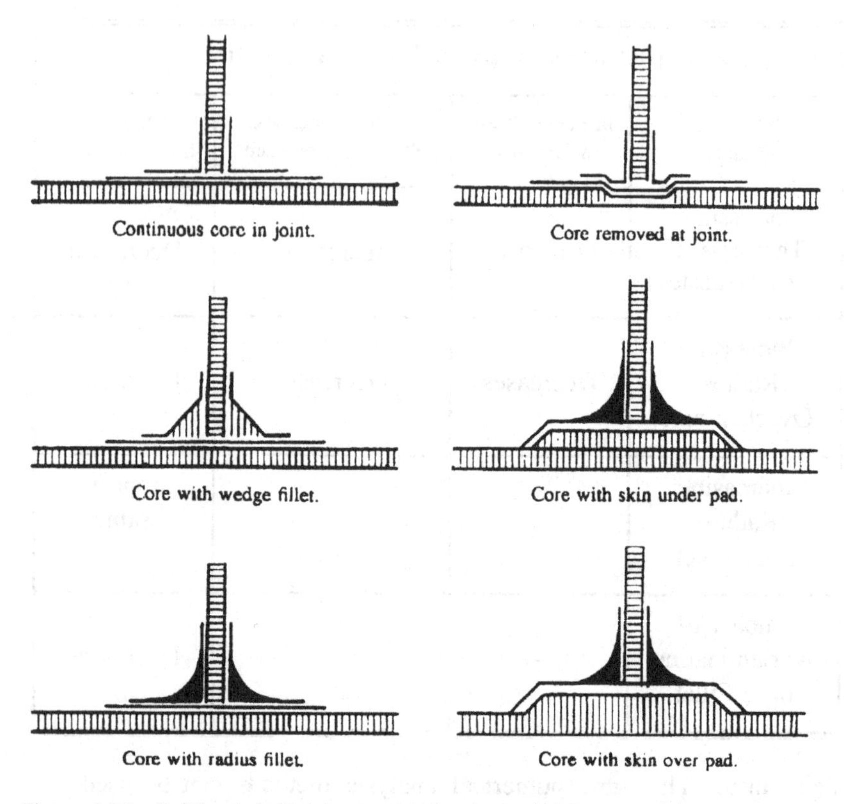

Figure 5.16. Bulkhead-shell joint geometries in sandwich construction.

marine applications the top hat stiffener is prevalent and is considered here. A typical arrangement is shown in figure 5.17. Typical design variables for such an intersection include:

a) Ratio of the intersecting stiffener heights
b) Thickness of external flanges
c) Length of external flanges
d) Radius of fillet
e) Fillet material
f) Number, position, and size of additional reinforcements

The two intersecting stiffeners should be of different sections, with the smaller one being continuous and the larger one being built over it. The greater the difference in height, the stronger the intersection will be. A ratio of at least 2:1 should be aimed for. This ensures that the tops of the stiffeners are not broken and so the global properties

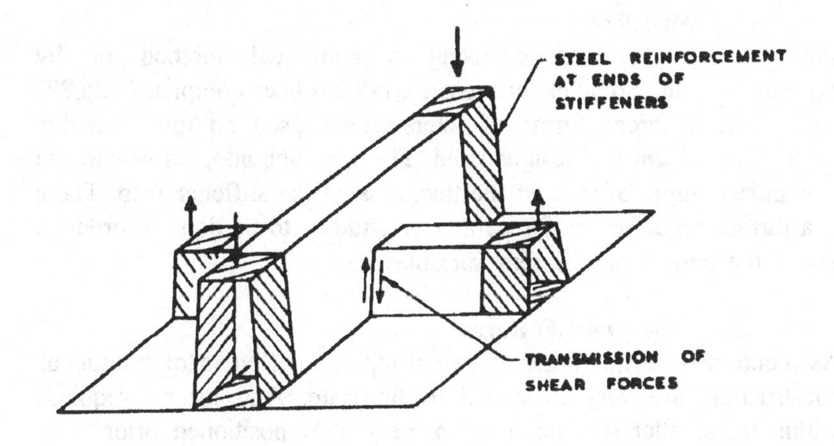

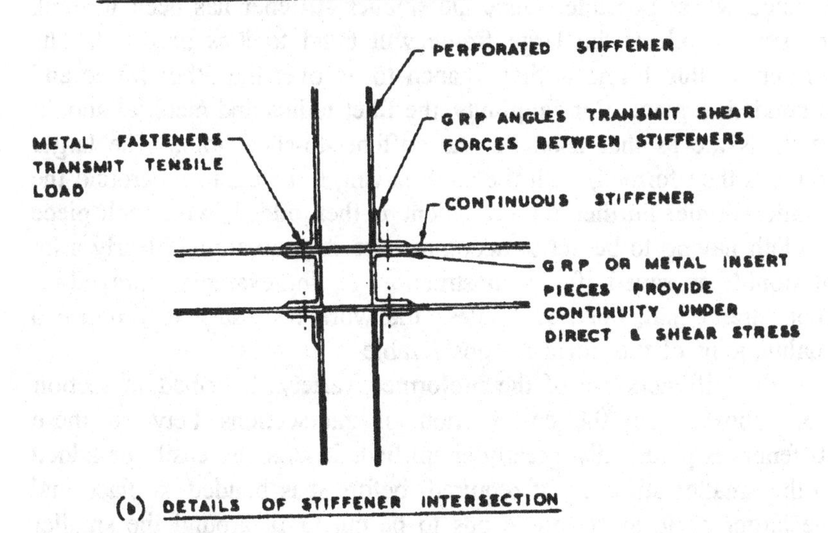

Figure 5.17. Typical arrangement of intersection between top hat stiffeners. (From [26]).

remain virtually unchanged. If this can not be arranged then insert pieces may be required as shown in figure 5.21. All that is then required is to attach the web of the large stiffener around the small one such that the direct, bending and shear loads are transmitted from one to the other. Since there is no access inside the larger stiffener this can only be done by external flanges.

5.5.3 Modelling

This can only be done using a numerical method or by experimentation. Some experimental work has been completed [26,27] on several different forms of intersections used on mine warfare vessels to confirm the designs used. The approach adopted was to use a boundary angle of at least the thickness of the stiffener web. There is a further requirement for parametric studies to be done in order to assess the impact of different variables.

5.5.4 Production Features

As mentioned earlier joints between top hat stiffeners of traditional construction are very awkward to fabricate. Any inserts required within the smaller stiffener must be accurately positioned prior to its fabrication. Any misalignment can affect the performance of the stiffener, particularly in compression; hence their use should be avoided where possible. Once the smaller stiffener has been formed, the surface where the large frame will bond to it is prepared. The former for this frame is then shaped to go over the other frame and is bonded in place. For simplicity the fillet radius and material should be the same as that used for the stiffener-to-shell joint. The larger frame is then formed, with the cloth having to be cut to go around the smaller frame. Further reinforcement is then added, with each piece of cloth having to be cut to accommodate the curvature. Clearly a lot of trouble is saved if the construction is, for example, sprayed-up chop strand mat, but nevertheless the work necessary to produce a quality joint of this form is considerable.

If the stiffeners are of the preformed variety described in section 5.3.1 above, then the construction of intersections between these stiffeners is potentially greatly simplified. Inserts can easily be added to the smaller stiffener, if required, before it is bonded in place and the larger preform simply needs to be cut to fit around the smaller frame with a suitable gap. The radius can be applied as before and the thin overlay requires much less work to form around the contours of the smaller stiffener. The suitability of this, as with any other, design needs to be confirmed, by either numerical analysis or experiments, for a given application.

5.6 REFERENCES

1] Smith, C.S., "Design of Marine Structures in Composite Materials", Elsevier Applied Science, London, 1990.
2] Godwin, E.W., Matthews, F.L., "A Review of the Strength of

Joints in Fibre Reinforced Plastics: Part 1 Mechanically Fastened Joints", Composites **11** (3), July 1980. p 155.

3] Matthews, F.L., Kilty, P.F., Godwin, E.W., "A Review of the Strength of Joints in Fibre Reinforced Plastics: Part 2. Adhesively Bonded Joints", Composites **13** (1), January 1982. p 29.

4] Matthews, F.L., "Joining of Composites" in Phillips, L.N., (ed.), *Advanced Composite Materials*, The Design Council, London, 1989.

5] Sadler, C.J., Barnard. A.J., "The Strength of Bolted/Riveted Joints in CFRP Laminates", Westland Helicopters Ltd, Structures Research Note No.16 1977.

6] ESDU Data Sheets Nos 85001, M2034, and E1048, "Program to Compute Stress and Strain Around Perimeter of Circular Holes in Orthotropic Plates".

7] Collings, T.A., "Experimentally Determined Strength of Mechanically Fastened Joints", in Matthews, F.L., (ed.), *Joining Fibre-Reinforced Plastics*, Elsevier Applied Science, Barking, Essex, 1987.

8] Benchekchou, B., White, R.G., "Influence of Stresses Around Fasteners in Composite Materials upon Fatigue Life in Flexure". Proc. AIAA/ASRE/ASCE/AHS/ASC Conf. *Structures, Structural Dynamics and Materials*, La Jollie, California, April 1993.

9] Grant, P.J., "Analysis of Adhesive Stresses in Bonded Joints". Proc. Symp. *Jointing in Fibre Reinforced Plastics*, Imperial College, IPC Press, London, 1978.

10] Corvelli, N. "Design of Bonded Joints in Composite Materials". Proc. Symp. *Welding, Bonding, and Fastening*, Williamsburg, Virginia, NASA TM-X-70269, 1972.

11] Hart-Smith, L.J., "Analysis and Design of Advanced Composite Bonded Joints". NASA CR-2218, April 1974.

12] Gibbs and Cox Inc., "Marine Design Manual for GRP", McGraw Hill Book Company, New York, 1960.

13] NES 140, "GRP Ship Design", Naval Engineering Standards, Issue 2, Undated.

14] "Rules for Yachts and Small Craft", Lloyds Register of Shipping, London, 1983.

15] "Rules for Building and Classing Reinforced Plastic Vessels", American Bureau of Shipping, New York, 1978.

16] "Rules for Classification of High Speed and Light Craft", Det

Norske Veritas, Hovik, 1991.

17] Dodkins, A.R., Shenoi, R.A., Hawkins, G.L., "Design of Joints and Attachments in FRP Ships' Structures", Proc. Charles Smith Memorial Conf. *Recent Developments in Structural Research*, DRA, Dunfermline, July, 1992.

18] Green, A.K., Bowyer, W.H., "The Testing and Analysis of Novel Top-Hat Stiffener Fabrication Methods for Use in GRP Ships", in Marshall, I.H., (ed.), *Composite Structures 1*, Elsevier, London, 1981.

19] Bashford, D.P., Green, A.K., Bowyer, W.H., "The Development of Improved Frame to Hull Bonds for GRP Ships - Phase 2", Final Report on MoD Contract NSM 7211/1011, July 1981.

20] Bashford, D.P., "The Development of Improved Frame to Hull Bonds for GRP Ships", Final Report on MoD Contract NSM 42A/0907, February 1986.

21] Clarke, M.A., "Stress Analysis of a Frame-Skin Connection in the MCMV". Internal Report, NCRE/L8/76, (Restricted), March 1976.

22] Hawkins, G.L., Holness, J.W., Dodkins, A.R., Shenoi, R.A., "The Strength of Bonded Tee-Joints in FRP Ships", Proc. Conf. *Fibre Reinforced Composites*, PRI, Newcastle-upon-Tyne, March 1992.

23] Shenoi, R.A., Hawkins, G.L., "Influence of Material and Geometry Variations on the Behaviour of Bonded Tee Connections in FRP Ships", Composites, **23** (5), September 1992.

24] Pattee, W.D., Reichard, R.P., "Hull-Bulkhead Joint Design in Cored RP Small Craft". Proc. 2nd Intl. Conf. *Marine Applications of Composite Materials*, Florida Institute of Technology, Melbourne, March 1988.

25] Shenoi, R.A., Violette, F.L.M., "A Study of Structural Composite Tee Joints in Small Boats", J. Comp. Mater., **24** (6), June 1990.

26] Smith, C.S., "Structural Problems in the Design of GRP Ships", Proc. Intl. Symp. *GRP Ship Construction*, RINA, London, October 1972.

27] Crabbe, D.R., Kerr, G.T., "Static and Short-Term Creep Tests on GRP Ship Type Connections for MCMV". NCRE/N236, March 1976.

6 PRODUCTION OF SHIPS WITH SINGLE SKIN STRUCTURES

6.1 INTRODUCTION

This Chapter explains aspects of single skin FRP boat and shipbuilding and covers a range of activities from facilities layout through vessel construction to post launch outfitting. The variety of materials and techniques available is immense, and so to cover the topics in the space available, emphasis has been placed on the requirements for larger vessel construction (in the 30 m to 50 m range) using more conventional materials, rather than high performance materials with an aerospace bias.

In some cases it is easier to describe concepts when applied to an example and to assist in this regard a vessel referred to as the "Case Study Ship" (CSS) is introduced here. Figure 6.1 shows a simple profile of the CSS which is a hypothetical vessel used here purely to help communicate some of the topics covered in this Chapter.

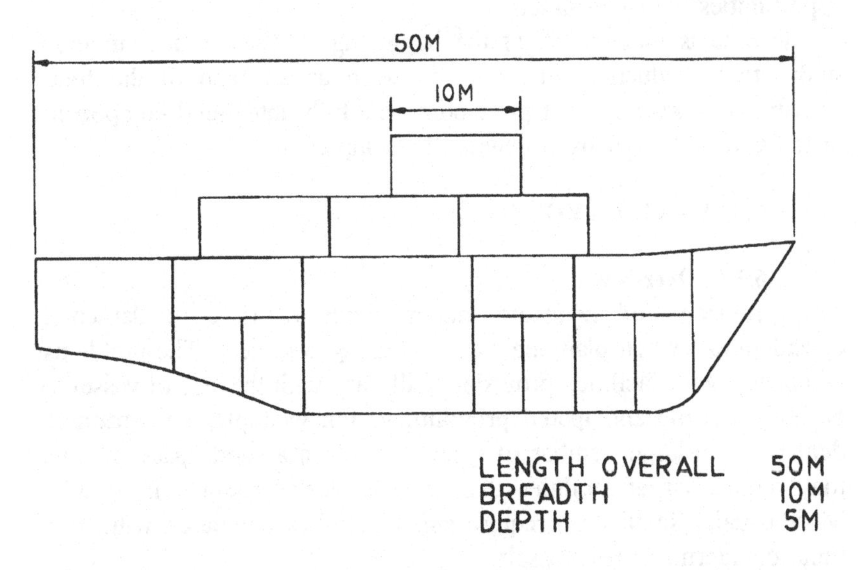

Figure 6.1. The Case Study Ship (CSS).

6.2 IS PRODUCTION THE END OF THE LINE?

Production is the end of line of the physical process of moving from a design concept to a fully functioning ship or boat.

Traditionally designs and even drawings were completed before the production team were able or allowed to impart their knowledge to the process. Production was, in that case, the "end of the line" in every sense.

It is crucial, however, that production requirements are considered at the very start of the project cycle. Any manufacturing task demands early consideration of the way in which the item will be produced. The purpose of emphasising the principal here is that with composites offering such flexibility in the production phase, thought must be given at a very early stage to extracting all the benefits before detailed design begins.

Production considerations should influence material selection and mould and tooling design. Early decisions have to be taken on construction techniques in order to design the structure to facilitate the best working practices. Composite materials offer tremendous potential for production friendly designs but unless this potential is realised and exploited from the outset there is a danger that opportunities will be missed.

The case is made here, at the beginning of this production case study, that production must not be seen as an "end of the line" activity; it is essential that production is a fully integrated component of the entire process from concept to completion.

6.3 YARD LAYOUT

6.3.1 Overview

The provision of adequate major facilities is a fundamental consideration when planning and evaluating a project. The problems associated with facilities provision will vary with the size of vessel to be built and the anticipated programme. For example, a programme demanding a large number of small vessels in a short space of time may require larger facilities than a single large vessel built slowly. More usually small craft require small facilities compared with, say, mine countermeasures vessels.

It is very unusual for entirely new facilities to be provided for a new vessel; it is far more common for existing facilities to be modified to facilitate the construction of a new design. This is normally due to funding and time constraints. The implication for production is that the 'ideal' production process can probably not be adopted due, for example, to cranage limitations, and constraints due to access and existing building heights. A compromise is therefore required which allows the vessel to be built in the most efficient manner available within the constraints. Early decisions should be taken on items such as panel and module size so that the design team can develop the ship accordingly.

6.3.2 Area Requirements

It has already been argued that area requirements are a function of vessel size and throughput requirements. An early agreement is therefore necessary to establish whether delivery is to dictate facility requirements or conversely facility constraints are to determine the delivery schedule.

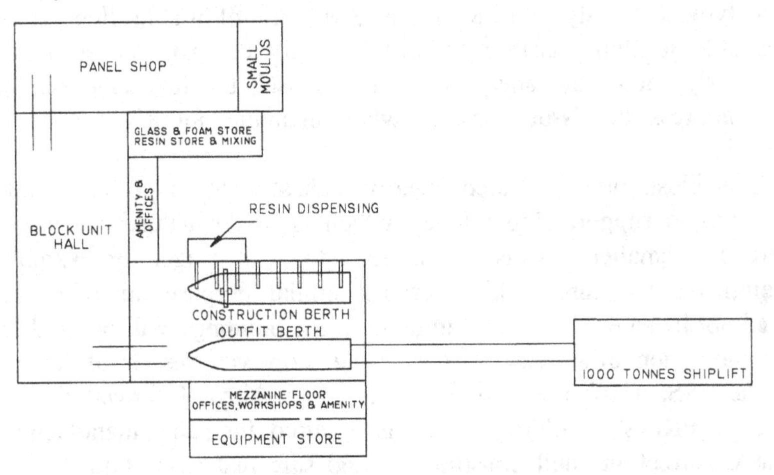

Figure 6.2. Plan of a typical GRP facility layout.

For the purpose of this case study it will be assumed that the facility shown in figure 6.2 is available and that six CSSs are to be built in this facility over a period of seven years. The planning assumptions are explained in section 6.3.3. Fortunately this facility has been designed to support the sort of programme which is required and is well provided with a Panel Shop, Block Unit Hall, Berth, Shiplift and buildings for supporting activities. Stores and manufacturing shops are located nearby and do not present a problem. There are adequate cranage and ship transfer facilities to allow for units weighing 30 tonnes to be lifted, a cross-transfer capacity in the berth of 300 tonnes and shiplift rail system of 1000 tonnes.

In terms which are simplified for the purpose of the case study the space requirements for one CSS are:

Hull	50 m x 10 m	=	500	m^2
1 Deck panels	5 x 10 m x 10 m	=	500	m^2
2 Deck panels	2 x 10 m x 9 m	=	180	m^2
Major bulkheads	5 x 10 m x 5 m	=	250	m^2
Superstructure units	4 x 10 m x 10 m	=	400	m^2
Hull internal units	3 x 10 m x 10 m	=	300	m^2
TOTAL			2,130	m^2

The typical facility illustrated has 2,900 m^2 of usable floor space available to ship construction (allowing for access). Whilst this is obviously more than adequate for one vessel, the following section will address the issues arising when planning for a sequence of vessels.

The illustration provided clearly indicates the type of facility required to support shipbuilding which is of the nature considered here. For smaller vessels in the 15 m to 30 m range, for example, facilities are required which serve a similar purpose but of course need not be so extensive and in many cases buildings will be used for a combination of activities. Considering, however, vessels of the size of the CSS, which are to be built in, say, a shipyard converted from steel to FRP shipbuilding, space is required for panel manufacture, unit construction, hull construction and safe materials storage. This example, albeit somewhat simplified, indicates the elements which must be accounted for.

6.3.3 Planning

Figure 6.3 shows a very basic programme for producing six CSSs over a period of just over seven years. It has been assumed that each vessel will take 12 months from commencement to the point at which it can be moved from the construction berth to the outfit berth. A further six months has been allowed at the outfit berth before the ship is launched on the shiplift. 10 months is allocated to completion work and trials. Each ship has an overall duration of 28 months and no learning curve has been applied.

The task of the planner will be to devise a sequence of component

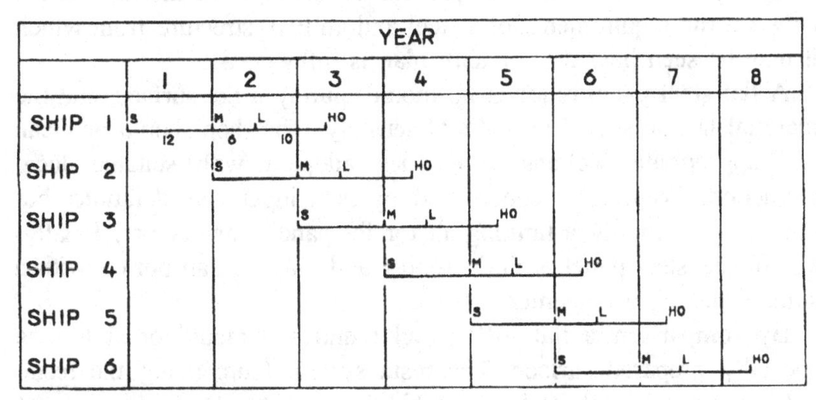

Figure 6.3. The CSS programme.

manufacture and assembly which will utilise the space available and meet the overall programme. The special features of the tasks which affect the plan will be raised as each is addressed in later sections.

The sort of capacity planning which is required at this stage is not unique to FRP ship or boat building and there are many common techniques which can be applied. FRP does, however, introduce some particular variables which can influence the plan and table 6.1 indicates factors present in FRP construction which are not considerations for steel fabricators.

Table 6.1. Factors influencing FRP capacity planning.
- Mould construction lead time
- Resin system shelf life
- Facilities restrictions due to safety and quality considerations
- Laminating sequences (upper limit on 'wet-on-wet' material)
- Curing times
- Transportation requirements for large components

A deck panel can be used to provide an example of the differences between the requirements of a steel and an FRP structure from which it may be seen how the capacity plan is influenced.

A flat steel panel requires no mould, purely a flat surface, and the material has no shelf life if stored sensibly. The shop should be clean and appropriate welding techniques adopted with suitable local extraction. Welding is constrained by heat input considerations but this is not normally a limiting factor for panel manufacture. Finally, a stiffened steel panel is fairly robust and may be transported with a simple bracing arrangement.

By comparison a flat FRP panel requires a mould or at least a specially prepared surface. The resin system (comprising the resin, accelerator and catalyst) has a shelf life which can be accommodated but requires some thought. Shop conditions for larger, high quality, panels should be such that heating and air flow are controlled and therefore it is normal to concentrate this activity in one location thereby reducing the options available to the planner. Large single skin component manufacture is rarely continuous due to curing requirements and this affects not only capacity planning but also resource planning as the work force often has to be employed on two or more different tasks during a shift. Finally, whilst FRP components cannot be considered to be fragile, they are prone to damage if components are lifted prior to incorporation into the intended structure without adequate support. Lifting apparatus is necessary for larger panels and these in turn require space and have an impact on lifting capacities.

FRP ships are not significantly more complicated to plan than steel ships, but there are differences and these should be recognised and accommodated from the outset.

6.3.4 Materials Storage

As with many of the variables in the facilities equation there is no simple all-embracing comment which can be applied to materials storage. Once again, it is important to consider the requirements of materials storage at an early stage, so that suitable locations can be identified.

Generally, reinforcements, core materials and foams should be stored in a dry environment, protected from high humidity at between 15°C and 20°C. Any moisture on any of these components which is present when laminating commences may have a detrimental effect on quality.

Accelerators, or promoters, may be stored under similar conditions to laminating resins, namely ambient temperatures, but ideally at approximately 20°C prior to use. Care should be taken to shield these materials from direct summer sun as the shelf life may be reduced as a result of heating.

Organic peroxides, which are commonly used as catalysts, should be stored under cool conditions (less than 20°C) and separately from cobalt accelerators. Direct contact between catalysts and accelerators is potentially extremely dangerous and must be avoided.

Pre-impregnated materials are normally cured by heating so such material is best kept at low temperatures. A large domestic freezer often provides an ideal storage facility.

Cleaning materials, of which acetone is a commonly used example, are highly flammable and must be treated as hazardous liquids. Aside from the safety aspects of its storage and use, acetone should be kept well clear of any laminate so that there is no possibility of contamination leading to a deterioration in quality.

All materials used for FRP construction should be stored in clean secure stores away from ignition sources. Care should always be taken to comply with manufacturer's recommendations, as well as current legislation, for example, the regulations governing highly flammable liquids in the United Kingdom. A point worthy of note here is that regulations governing the storage and use of hazardous materials are constantly under review and countries in the EEC could experience changes in this area in the near future. Users of these materials should ensure that they have a mechanism in place to provide updates of relevant regulations.

6.4 CONSTRUCTION AND OUTFIT BERTH

6.4.1 Layout

The sketch of a typical berth layout provided in figure 6.2 indicates a construction and an outfit berth. The construction berth provides a facility to produce hulls in a location which is served by the necessary materials handling equipment and ventilation. A dedicated area saves duplicating expensive facilities and also reduces the amount of mould handling required. The hulls can be transferred sideways using a rail system to the adjacent outfit berth where open-sky outfitting will be followed by final assembly of structural components.

6.4.2 Hull Moulds

Several of the factors affecting the design of the hull mould and the choice of material are considered below:

6.4.2.1 Vessel size

The size of the vessel will be a factor in determining the way in which the mould is sectioned to facilitate de-moulding. Moulds for small vessels are commonly produced from wood or FRP whilst for larger vessels aluminium or steel is used. Large moulds built in steel become very heavy and difficult to move, so for that reason aluminium is often used although this is more expensive.

Occasionally jigs rather than moulds are used to create a hull shape. In this instance the bow section is moulded conventionally, but the remainder of the vessel is produced from thin flat panels set into the jig to create the outer surface of the hull. Laminate is applied to the inner face of the thin panels as it would be to a mould. This method has been used to facilitate production of vessels of different length from the same basic jig.

6.4.2.2 Required durability

Naturally a mould for a single, never-to-be-repeated vessel, will not be required to be as robust as a mould for a series of vessels. In the former case a much cheaper more simple solution could be adopted.

Durability is also influenced by the storage conditions of the mould between uses. A wooden mould stored in the open will not fair so well as a properly coated steel mould.

6.4.2.3 Available skills

One particular yard may have the skills available to build a mould from timber, whilst another may not. This apparently minor consideration may influence the design of the mould and is therefore worthy of discussion in the early stages.

6.4.2.4 Production processes

A fundamental decision to take is whether a female mould or male plug is to be used to create the hull. Single skin structures are commonly built in moulds, one great benefit of which is the quality of finish on the outer surface of the hull. There are examples of single skin hulls built on plugs, one of which was videoed in the Philippines where consolidation was done by the bare feet of the laminators. Quality of the outer surface finish was perhaps not an issue in this instance!

There are many other ways in which production methods may affect the design of the mould. Material selection is a factor since this will determine whether the resin is to be applied in liquid form or pre-impregnated into the cloth. If vacuum techniques are to be employed, a different mould may be required from that needed for hand lay-up. The variations are enormous, highlighting again the need for clear, careful planning of the process to avoid costly changes later in the production sequence.

6.4.2.5 The CSS mould

Having considered the options it has been decided that the hull mould for the CSS will be manufactured from mild steel. The justification for this includes consideration of the size of the mould, its required life span, general durability, the fact that it will remain in a covered area, and since the sections will be designed to roll away from the hull and part of the mould will be stationary there are few segments to be lifted away by crane. Had mould weight been a limiting factor aluminium would be used for its construction. Figure 6.4 shows the type of mould envisaged.

Figure 6.4. The mild steel mould.

The mould will also be designed to support a portable gantry to provide the laminators with a safe working platform.

6.4.3 Material Dispensing and Consolidation

The choice of material for the hull will be determined, or at least the range narrowed, by the design criteria of the structure and cost constraints. The final choice of material should only be made after production trials have shown that the selected combination will actually perform as expected when it is used by the people who are going to do the job. The skill and experience of the people using the materials is a significant factor in FRP manufacture. Three methods of producing a single skin hull are considered separately in the following sub-sections.

6.4.3.1 Hull construction using 'prepregs'

'Prepreg' is a term commonly used for reinforcements impregnated with resin (very often an epoxy resin) which are supplied with peelable film. The resin is cured at reasonably high temperatures, 80°C to 120°C is not uncommon, and to improve shelf life these materials are stored below 0°C.

Prepregs are generally applied to a plug and covered with suitable material to enable a vacuum to be applied during curing. The entire assembly is then placed in an enclosure and heat applied for the desired period.

Prepregs are associated with high performance materials and are generally significantly more expensive than a more traditional glass and polyester matrix. The cloths are almost always applied to the plug or mould by hand, which is a quite straightforward process since prepregs taken from a cold store are not too tacky and emit very little odour. Prepregs allow very high levels of quality to be achieved leaving very little to the discretion of the laminator. In these days of heightened safety awareness the low fume emission levels associated with this process are attractive.

The main disadvantages with the use of prepregs for large hulls are the cost of the material and the difficulties, and again expense, of heating and applying vacuum to large areas. Despite these disadvantages the use of prepregs can afford significant advantages for single skin hull construction on the right applications.

6.4.3.2 Hull construction using "wet laminating"

The term "wet laminating" is used here to refer to the application of a liquid resin to a glass or other reinforcing fabric. The most common form used for larger vessel construction is a woven glass fabric to which a polyester resin is applied. There is a wide range of materials within these this group and, when considered with the options on catalysts and accelerators, a very large number of combinations is possible.

It is important to focus on a small number of possible glass/resin combinations to preserve research and development resources and use them to their best advantage. The selection of the process is crucial and significant; within this is the decision whether or not to apply the materials by hand or machine. Mechanised material lay-up allows the use of very much heavier cloths than is practical by hand. However, heavier cloths exhibit different "wetting-out" characteristics from lighter materials and it would be dangerous to assume that results from trials on lightweight fabrics could be simply extrapolated to their heavier brethren. Materials must be selected and tried and tested in exactly the same way as they will be used for production.

To utilise machines to apply this horribly sticky, uncooperative mass of wet laminate to a mould seems a very attractive proposition. If it works and if it is cost effective this is true, but this is not a decision which should be taken lightly.

It is assumed that the hull of the CSS is to be produced from a polyester/resin/glass fibre matrix and the quality required has to be at least equal to that required by a Classification Society. The quality of the laminate must be consistent and it will be important to control the relative proportions of resin and glass and exclude air from between the layers of material.

Experience has shown that it is quite easy to use a machine to "wet-out" glass cloth with resin. It is not so easy, however, to use a machine in a hull mould when air inclusion must to a degree be avoided and where there is a reasonable amount of shape in the vessel. This is simply because the machine is best at long continuous runs which is not compatible with the constant adjustment and tailoring needed to accommodate the shape of the vessel, and the care required to exclude air from the laminate. Single skin hull construction using wet laminates is not impossible to mechanise, but experience shows that in order to achieve consistently high quality it is very difficult for the machines currently available to beat manual methods. Laminating machines are not able to cope with stop/start operations.

The hulls of the series of CSSs will not be built using a machine which impregnates the cloth and offers it up to the hull mould, because it has been adjudged that there will not be sufficient continuous running to justify such a machine. Time and effort will therefore be expended on providing good access to the place of work so that the laminators can operate safely and in an unhindered way. Materials handling will be considered in more depth later but it is relevant to mention here that resin will be dispensed at a point close to the hull mould ready for manual application and adequate provision has been made for accepting deliveries of cloth.

A further option is available to mechanise, at least in part, the application of materials to the mould surface. If the reinforcement is to take a chopped strand rather than a woven form, the resin and strands could be applied using a gun which sprays resin and short glass strands onto the mould. Even if the reinforcement is woven or knitted, the resin could be applied using a spray gun. Both methods have been used for suitable applications but not commonly to produce higher quality laminates because of the inherent difficulties with controlling material thicknesses and glass/resin ratios. The control of styrene vapour in the atmosphere is a problem which is only exacerbated by using resin spray guns for large jobs in an open environment.

When the hull plating has been completed to the required thickness, frames and longitudinal stiffeners are built into the structure. These are normally created by bonding onto the hull plating foam formers or thin FRP formers over which the required thickness of laminate is laid.

Following completion of the frames and longitudinals, the bulkheads, decks and initial units can be installed. When there is sufficient structure to hold the hull firm the mould can be removed from the ship in a planned sequence. Alternatively, if the facility has adequate lifting capacity, the hull can be lifted from the mould reducing the amount of disruption between successive mouldings. Figure 6.5 shows this method adopted for a 30 metre hull clearly showing the primary structure.

In the case of the CSS the port side of the mould will be removed and selected sections along the keel. The weight of the hull will be supported on trollies which run on a rail system. The ship will be released from the remainder of the mould by injecting air between the mould surface and the hull. Once free, the hull is moved sideways to the adjacent berth from where outfitting and structural assembly will continue.

6.4.3.3 Hull construction using resin transfer moulding

Whilst not yet applied to large hulls, there is considerable potential

Figure 6.5. Lifting of hull from the mould.

for the application of resin transfer moulding processes to such tasks. Resin transfer moulding (RTM) may be carried out by injecting resin under pressure, into an enclosed space or mould assembly. The expense of manufacturing matched moulds for limited numbers of very large items is quite obviously prohibitive. There is a variant of RTM, however, in which the resin is introduced under a vacuum and does not require a double sided mould.

One such process is the Seemann Composites Resin Injection Moulding Process (SCRIMP) which has been developed in the United States of America, and is licensed to Vosper Thornycroft (UK) Ltd for the United Kingdom and European Economic Community. This process has the advantage of being easily used with existing moulds and requires little in the way of special equipment. In very basic terms the laminate is prepared in a dry form with all the reinforcements arranged in situ. This is covered with an air-tight membrane which is sealed at the ends and a vacuum is applied to evacuate the air from the space enclosed by the cover. Resin is allowed to flow into the reinforcements and in so doing wets-out the fibres which are consolidated by the membrane.

The SCRIMP process may be applied to a variety of tasks including hull and panel manufacture, and development work is rapidly increasing the size of component which can be moulded.

The ease with which existing moulds can be adapted to SCRIMP has already been mentioned. Further benefits include the high quality of laminate since there are few or no air voids, and very much improved safety levels because the resin is not exposed to the atmosphere whilst curing.

It is not recommended here that SCRIMP is applied to the CSS because it is recognised that there is development work required to apply the process to very large FRP structures. The whole concept of RTM in its various forms has great potential and in future years it will supersede many of the laminating methods currently in use.

6.4.4 Productivity

Previous mention has been made of the fact that mechanisation of hull construction does not necessarily result in high productivity levels. Productivity can be improved by providing good access to the job, a safe environment and the right materials and the right time. FRP construction is not unique in this respect.

Anyone who has worked with a wet laminate will know how uncooperative it is. It falls off overhead surfaces, it slumps down vertical faces and it sticks to scissors and hands. Productivity can readily be increased by reducing overhead laminating, either by rotating the job or the mould and eliminating wherever possible the requirement to cut and tailor wet cloth.

Smaller hull moulds can be rotated to provide good access to the workpiece, but this becomes prohibitively expensive for larger craft. Where this is the case, higher productivity can be attained by keeping the process as open and uncluttered as possible so that long lengths of cloth can be put down with the minimum of hinderance. If more complicated areas are necessary, it is often better to build them separately rather than trying to undertake difficult tailoring at the end of each long cloth and consequently slowing the whole process down.

Stiffeners with many intersections are expensive to produce; so again productivity can be increased by keeping the design as simple as possible, by for example, designing a longitudinal stiffened structure rather than a grillage.

6.4.5 Ventilation and Safety

When resin cures, the solvent, which is a large proportion of the liquid, evaporates. Polyester resins release styrene vapour and the curing process is aided by providing an air flow over the surface of the laminate to prevent saturation of the immediate atmosphere. This in itself helps ventilate the workplace but may not be sufficient to provide a safe working environment.

The scale of the operation will determine the requirement for forced ventilation. In fact some smaller shops using small amounts of material need little or no specialist equipment. Larger operations such as the 50 m CSS will require mechanical ventilation to ensure that styrene concentrations are maintained below the prescribed maximum. In the United Kingdom the current maximum exposure level for styrene in the atmosphere is a time weighted average of 100 parts per million. At the height of a hull lay-up, with a lot of wet material around, a good ventilation system will keep the styrene proportion down to about 70 ppm. It is important to have a method of regularly monitoring the styrene concentration and a variety of methods are available ranging from simple hand held devices to sophisticated air sampling units using spectral analysis. It is quite wrong to plan to work in areas of high styrene or vapour concentration, but it is occasionally necessary to finish work after the level of styrene has become unacceptably high and on such occasions face masks with suitable organic vapour filters should be used. Laminators working in enclosed spaces such as tanks should wear positive pressure air-fed breathing apparatus. The regulations governing operations in environments such as FRP facilities differ between countries and in some cases between regions. It is generally expected that exposure limits will be tightened in the near future and fabricators should be planning for this eventuality.

Other sensible necessary precautions which should be adopted when working with liquid resins include complying with flammable liquid regulations such that ignition sources are kept at a safe distance and electrical equipment is either safe by distance or adequately shielded to prevent sparking. Suitable protective overalls should be worn by laminators and it is normal to use a barrier cream to protect the hands. Eye protection should be used when dealing with catalysts and when there is any danger of liquids splashing into the eyes.

6.5 PANEL SHOP

6.5.1 Layout

When building smaller craft, a dedicated Panel Shop is not a requirement and very often deck and bulkhead panels can be adequately built in an area of the berth. A Panel Shop is necessary when large panels are to be built in significant quantities to support the construction of a series of ships. The panel facility is simply a number of flat or cambered moulds on which plating can be produced, which is subsequently cut to create structural components. The Panel Shop in figure 6.2 has four 18 metre x 12 metre panel moulds and two 12 metre x 12 metre panel transfer spaces. If panels can be turned elsewhere (in a Block Unit Hall for example, see figure 6.6) this facility does not need to be particularly high.

Figure 6.6. The block unit hall.

6.5.2 Panel Moulds

Panel moulds can be fairly simple since the aim is normally to produce a flat plate of a given size. There may be occasions when a deck with a camber or a knuckle is required, even so the mould can be produced using the same principles.

Typically the base of the mould will be a heavy steel or wooden framework, sturdy enough to be walked on without damage and able to withstand warping. The base may be covered with blockboard or plywood, the thicknesses of which will depend on the robustness required and the spans between beams. The laminating surface is best formed by overlaying the blockboard with plywood faced with melamine or a similar material. Melamine provides a robust mould surface which requires little maintenance. It is worth taking a good deal of care with the finish of the joints and filling of screw holes because every blemish will be reproduced on each FRP panel produced.

Large panels will require points set into the surface so that jacks can be used to free the panels from the mould. Releasing panels without jacking or similar assistance can be extremely difficult.

6.5.3 Material Dispensing and Consolidation

Panel manufacture comprises two main operations: plating production and stiffener fabrication. Panel plating is the easiest process in FRP ship construction to mechanise; it is also the most productive operation undertaken by hand. Stiffener production is a great consumer of man hours but is by nature difficult to apply machines to. It is also much easier to apply machines to panel plating manufacture than to hull plating, but in order to justify the expense of a machine for material dispensing, one must be sure of it outstripping the very high productivity which can be achieved by hand.

Machines are available on the market which will satisfactorily wet-out glass cloth with polyester resin and present it to the mould surface. Consolidation may be undertaken by laminators using rollers whilst standing on the dispenser gantry. Automatic consolidation using a series of rollers fixed to the gantry is possible, but is not in widespread use in applications requiring a high quality laminate with a low air void content. Dispensers for panel plating lay-up can be slung under crane gantries or fixed to a dedicated gantry which runs on rails either side of the panel mould.

Stiffener manufacture is far more difficult to mechanise since stiffeners are rarely uniform in cross-section, and even where they are there is often a requirement to increase the widths of successive cloths to achieve an overlap on each set of "tails". Any tailoring at stiffener ends or intersections adds further complication. These difficulties often result in stiffeners being produced manually.

The Panel Shop shown in figure 6.2 has been configured to produce large panels to support production of the CSS. In this case panel plating will be produced by two simple dispensers each on its own dedicated gantry. The dispensers will each carry large rolls of cloth and will be fed with resin pumped directly from the bulk resin store. The dispenser catalyses the resin immediately prior to use. The cloth is fed through a shallow resin bath on the machine before dropping down to the mould surface. Plating consolidation will be done by hand in order to be certain of consistently achieving the quality of laminate required. Stiffeners will be laid-up by hand over foam or FRP formers.

6.5.4 Productivity

Productivity is affected by similar factors in both panel and hull lay-up, in the sense that large uncomplicated areas are far cheaper to produce than those which involve a degree of complexity. There is scope to increase productivity with panel plating as the panels cure horizontally a less viscous resin can be utilised compared with the relatively thixotropic resins required for hull lay-up. A combination of "thin" resin and dispenser can facilitate the use of very heavy cloths, meaning that large quantities of material can be deposited in each run along the panel. As a consequence, panel plating manufacture can be more than four times as productive as hull plating.

Whilst minor stiffeners can be pre-made sections bonded to panel plating this is generally not practical for main structure. Stiffener fabrication remains a relatively labour intensive activity and normally accounts for three quarters of the time required to produce an entire deck panel.

6.5.5 Ventilation and Safety

Large panel moulds require ventilation ducts to be arranged to move air across the surface of the panel to assist curing. Smaller facilities may not require any special arrangements.

Providing a suitable resin can be found, panel manufacture is an ideal application for a low styrene emission (LSE) resin. An LSE resin can reduce the workshop styrene level by 50% compared with a standard resin during the curing phase. Care must be taken to ensure that the LSE resin provides the mechanical properties required in the final laminate.

Other measures for personal safety are similar to those required for hull laminating.

6.6 BLOCK UNIT HALL

6.6.1 Layout

The Block Unit Hall is essentially a large space to allow structure to be set out for pre-erection outfitting and the subsequent assembly of panels to create three-dimensional units. The Block Unit Hall must be large enough to provide safe access around the panels and units, and also to accept the constant flow of materials required to support activities in the area. A typical view inside such a building is shown in figure 6.6.

Each workstation inside the hall must be served with adequate services to support the operations planned to take place there. The example in figure 6.6 shows the services provided from overhead booms which can be pulled aside to facilitate panel and unit handling. Typically the services provided will include air supply and extraction, compressed air, dust extraction and styrene monitoring equipment.

6.6.2 Pre-Erection Outfitting

Panels produced in the Panel Shop will be complete structurally, in that all the laminating will be complete. Once at this stage they will be transferred to the Block Unit Hall, still inverted, where all brackets, cable supports, pipe supports and other preliminary fittings will be attached. Care must be taken to keep items clear of areas on which boundary angles will be laminated at a later stage. Panels will be painted and insulated next after which minor cables, minor items of equipment and pipes can be installed.

Figure 6.7. Pre-erection outfitting on panel.

Figure 6.7 shows a panel which has been brought up to this stage and is in the process of being turned so that outfitting can continue on the top. Two significant features are apparent in figure 6.7. Firstly a very robust lifting and turning frame is required to hold panels to prevent damage during this operation. Secondly the turning process is a significant factor in determining the height of a building such as the Block Unit Hall. A 10 metre deck panel requires at least 13 metres under the crane hooks to enable it to be turned safely.

A key to successful advanced outfitting is early definition of the vessel and a controlled flow of materials and equipment. Traditionally compartments such as the bridge would be defined late in the design and drawing sequence. Pre-erection outfitting may, for example, demand information for seating on the bridge deck relatively early in the programme and this can have a significant effect on the way the drawing programme and materials scheduling is approached.

It is important to appreciate that the substantial benefits to be gained by adopting principles such as block construction and advanced outfitting do not happen by accident. The very earliest stages of design must make provision for the way in which the vessel is to be constructed. Thought must be given to the block sizes, what they contain, and how they interrelate to each other. It is advantageous for example to create a block with compartments containing electronics equipment, and another which is predominantly accommodation spaces. This approach not only serves to reduce the requirement for services (cables and pipes connecting compartments for example), it also enables the most important areas of the vessel to be worked on in isolation from those which are less important and possibly have a lower work content.

6.6.3 Block Construction

After the panels have been outfitted on both sides they are gathered together for assembly into three-dimensional blocks. The view of the facility in figure 6.6 shows this activity at various stages. Some of the blocks may be built upside down to avoid large quantities of overhead laminating. In such cases a turning frame such as that used for panels is required so that semi-finished blocks can be rotated in order to be completed.

Block construction is the assembly of a kit of parts bonded together with boundary angles. One of the most critical features of this stage of ship construction is ensuring that the components are in the correct position relative to each other and are held rigid during laminating. A number of fairly simple jigs can be used to hold components in place, and it is important to check nominated dimensions regularly to monitor and correct any movement which may take place.

The CSS is designed to be readily adaptable to the principles of block construction and, as such, has three internal and four superstructure blocks. The internal blocks are generally an assembly of decks and bulkheads which form tanks or small compartments in the bottom of the ship. The great advantage of building these structures separately from the hull is that the majority of laminating can be done in the Block Unit Hall with the block inverted. The benefit of this method compared to the alternative of laminating overhead in a confined space using breathing apparatus can be readily appreciated.

When the blocks are complete with all internal structure and minor bulkheads fully installed, outfitting is continued to ensure that the maximum amount of equipment and systems are fitted prior to taking the block to the hull. Figure 6.8 shows a vessel where the bridge unit has just been lifted into position on top of the other, previously erected, superstructure units.

Figure 6.8. Installation of bridge unit.

6.6.4 Outfitting Techniques

Brief mention is made here of some of the outfitting techniques applied to single skin structures, partly to indicate the relative simplicity of the methods employed and partly to enable the reader to identify whether sandwich structures require alternative techniques.

Since the majority of single skin structures are relatively thick, compared to the skin of a sandwich structure, this can be used to good advantage when fitting brackets and supports. Minor items such as single cables or light switches, for example, can be easily fixed to bases bonded onto the structure. Heavier items such as electrical equipment or bookcases may be fixed by bolting directly through the deck or bulkhead. It is not usual to tap holes into FRP since the thread tends to strip easily, however, there is a varied range of metal inserts on the market which will accept a bolt or machine screw.

The techniques used will vary between types of vessel, with smaller craft generally not having heavy equipment and therefore the emphasis being more on bonding rather than bolting equipment and fittings. Securing items, even on a minehunter which could be subject to high shock loadings, is not a major problem and in most cases a solution can be found by a fairly straightforward bolting arrangement.

6.7 POST LAUNCH OUTFIT

If facilities and programme constraints allow, all laminating and the vast majority of outfitting should be completed in the berth before the ship is exposed to the elements. The benefits of this approach arise because the work is not weather dependent and usually the bulk of the facilities are in close proximity to the berth.

Post launch outfitting should therefore be confined to the installation of minor systems, connecting equipment and setting-to-work programmes. These activities are not peculiar to FRP vessels and each ship will go through this phase to some extent depending upon the degree of complexity of the equipment fitted.

6.8 MATERIALS HANDLING

6.8.1 Resin Systems

The term 'resin systems' is used to include liquid resin and the associated catalysts and accelerators necessary to make the resin cure.

Reference has already been made to the safety requirements for the storage of these materials and the need to comply with both statutory and manufacturer's guidelines. It is important, if any doubt exists, that expert advice is sought to ensure the safe use and storage of resins, catalysts and accelerators.

It is possible to vary the way in which resin is stored and distributed to suit the scale of production and the techniques employed. A small yard using hand lay-up or spray application methods may find it convenient to take resin in 45 gallon drums and simply add accelerator and catalyst to small quantities when drawn.

Larger operations can opt to take resin in 1.5 tonne tanks which can either be connected directly to a machine or decanted into smaller containers for distribution to hand lay-up squads.

Facilities using high volumes of resin, and this would include the CSS facility, would be served by large resin silos (usually two) which are charged from 20 tonne capacity road tankers. A benefit of having two silos is that backup is provided in the case of one silo becoming contaminated. Resin is distributed from the silos by a pipe system to the Panel Shop for both hand and machine lay-up and to the Block Unit Hall and Berth for hand lay-up. Bulk resin is supplied in a pre-accelerated form.

The resin, which is deposited by machine, is catalysed as it enters the resin bath just prior to wetting-out the fabric. The machine should be calibrated to ensure that the catalyst proportion is correct and the right quantity of resin is being applied to the cloth.

Resin used for hand laminating will be drawn locally from the distribution system, if piped, or from a central issuing location. At this point the material should be weighed and also a check carried out to ensure gelation will occur. A 'gel-test' can be completed in about two minutes by forcing the gelation of resin in a test tube held in boiling water.

Whichever method is chosen for resin distribution the equipment will become very messy, very quickly. Good housekeeping is therefore essential to prevent contamination and seizing of all moving parts.

Pre-accelerated resin is convenient since accelerator does not need to be stored separately and the task of measuring out and adding accelerator to resin becomes superfluous. However, it may be necessary to hold stocks of accelerator for a variety of reasons, and where this is the case it can be stored with the resin in the containers in which it is delivered. Resin can be stored at ambient temperatures, so long as it does not become too hot in summer (warmer than 30°C) and in winter it is warmed to about 20°C before use.

Catalysts, or organic peroxides, must be stored away from accelerators in cool conditions. Catalysts should generally not be subject to temperatures above 20°C. Manufacturer's recommendations must be followed.

6.8.2 Core Materials and Reinforcements

The storage and handling requirements of foam, balsa and reinforcements are relatively straightforward. They should be stored in a clean area which is warm (say 15°C in winter) and dry. It is essential that local conditions are taken into account to ensure that during storage and transfer to the laminating area these materials do not become damp nor should condensation be allowed to form on them.

Large rolls of fabric can go directly to machines, but materials required for hand laminating are ideally cut to size and marshalled into a kit of parts so that the laminator can apply them to the job with the minimum of preparation at site.

6.8.3 Cleaning Materials

There is a range of cleaning materials available on the market, some of which are very volatile and have low flash points. In some cases these are the most hazardous materials stored and handled in the facility in terms of potential flammability. To try to give rules here for their use and storage may be dangerously misleading. It is important to recognise the need to use certain liquids carefully and in a way which complies with the requirements of any regulations in force.

6.9 QUALITY ASSURANCE

6.9.1 Training

The requirements for quality assurance will vary with the type of work being undertaken and the conditions imposed by the customer. It is important to get this balance right so that appropriate quality measures are taken and the people carrying out the work fully understand what is expected of them. No one wants a job to be rejected because it is sub-standard; alternatively it is not cost-effective to produce a laminate which is far superior for its intended use.

Most medium size and large companies will have an "in-house" quality department which develops and oversees quality policy. In many cases customers will supplement this by introducing their own quality inspectors or will appoint a third party to act on their behalf. Before a project commences, all interested parties must agree on the standards which are to be applied to the job. These standards must of course be achievable in practice.

A fundamental starting point on the road to the required quality of product is laminator training. This should include all the primary tasks which the laminator will require in the production phase, and test pieces should be evaluated against pre-determined test criteria. Particular importance should be attached to consistently achieving the standards for air exclusion and glass-to-resin ratios.

6.9.2 Materials Inspection

Inspection should be applied to the receipt of materials to ensure that certificates of conformity are provided where necessary. Stock rotation and shelf lives are not normally the responsibility of quality inspectors, but they should carry out audits to check that satisfactory procedures are in force.

The quality assurance team has a role in monitoring material issues. This is less of a task for reinforcements than it is at resin issuing points. The consequences of resin being issued and used which is either un-catalysed or inappropriately catalysed may be very severe. Tight controls applied to this part of the process are never wasted and provide reassurance to both the company and customer alike.

During the laminating process audits are common at all phases. Areas which are critical to the laminate will receive more attention than others and examples of these are: boundary angle radii, laminate thickness and void content. Laminates built to withstand particularly arduous conditions, minehunter structures for example, will be governed by a range of procedures which are audited regularly during the working day.

Completed and part completed laminates are often tested using a variety of destructive and non-destructive methods. Destructive test samples are taken from openings for hatches or similar areas within panels and may be subjected to interlaminar shear tests or ashing tests, to establish the glass-to-resin ratio, for example. Non-destructive methods include shining a light through the laminate (if it is not pigmented) to inspect for voids and thickness measurement using ultrasonic equipment.

6.9.3 Shop Conditions

Quality can be adversely affected by sub-standard materials and ill-trained people, but even if these factors are satisfactory the laminate can be ruined by poor shop conditions. Although it is less the case now, FRP had a poor reputation, certainly in the pleasure boat sector, for sub-standard products. This could be attributed in part to inappropriate shop conditions.

Temperature is an important variable for all resin systems. Resins are available to meet a range of ambient conditions and naturally prepregs require special consideration. Shop temperatures should be maintained within the specified temperature range (commonly 16°C to 25°C) and laminating should cease if temperatures outside the limits are experienced.

Another feature which is detrimental to quality is high humidity. If temperatures can be maintained above the dew point, few problems should be experienced. Moisture in the laminate must be avoided, and care should be taken to ensure that by moving reinforcements from a store to a workshop no condensation has formed on the fibres.

Finally, the laminate should be clean. Cutting composites is a dusty process, requiring dust extraction to be used in the majority of cases. Dust reduction is necessary to stop inhalation and appropriate masks should be worn by operatives. The dangers of working in dusty environments are widely appreciated now. Dust should also be minimised as it can seriously contaminate laminates resulting in significant reductions in physical properties. Suitable dust extraction equipment should be used when cutting or grinding and a regular shop cleaning routine adopted.

6.10 BIBLIOGRAPHY

1] Chalmers, D.W., Osborn, R. J., Bunney, A., "Hull Construction of MCMVs in the United Kingdom", Proc. Intl. Symp. *Mine Warfare Vessels and Systems*, RINA, London, June 1984.

2] Bunney, A. "The Application of GRP to Ship Construction", Trans NECIES, **103**, (4), 1987.

3] Naval Forces Special Supplements for Vosper Thornycroft (UK) Ltd., Mönch, Farnborough, 1989/1991.

4] Critchfield, M.O., Judy, T.D., Kurzweil, A.D., "Low-Cost Design and Fabrication of Composite Ship Structures", Proc. Charles Smith Memorial Conf. *Recent Developments in Structural Research*, DRA, Dunfermline, July 1992.

5] Johansson, M., "Styrene Emission during the Moulding Operation, Reinf. Pl., October 1991.

6] Dixon, R.H., Ramsey, B.W., Usher, P.J., "Design and Build of the GRP Hull for HMS Wilton", Proc. Intl. Symp. *GRP Ship Construction*, RINA, London, October 1972.

7] Harris A.J., "The Hunt Class Mine Countermeasures Vessels", Trans. RINA, **122**, 1980.

7 PRODUCTION OF YACHT HULLS OF SANDWICH CONFIGURATION

7.1 INTRODUCTION

In this Chapter a production case study of Admiral's Cup type of sailboats is shown. The sandwich construction is made out of advanced composites. Reflecting the high costs of these materials new production methods have had to be used to ensure that the requirements concerning specific properties are fulfilled.

7.2 YARD LAYOUT

7.2.1 CIM Orientated Production

Although computer integrated manufacturing has hardly been installed and used in shipyards, CAD generated designs and templates are standard, even for the smallest vessels.

That means two-dimensional parts like ribs and webs are sized with the help of templates, blue prints or 2-axis cutting machines.

For series boat production complete moulds - either male or female - are 5-axis machined up to a size of 15 m length and 5 m width to achieve ultimate precision and performance.

Within the layout of the boatyard there is a need for a CAD room, if possible an NC programming room for multi-axis work and within the shops, a multi-axis machining centre. (See figure 7.1 and figure 7.2).

7.2.2 Need for Advanced Composites

Weight reduction in boatbuilding is realised through the usage of advanced fibres and cores coupled with production methods achieving the best possible specific mechanical properties. Pre-impregnated fabrics and tapes (prepregs) are the key elements for this requirement and are now widely used.

Cooling rooms down to -16°C and ovens or autoclaves up to 120°C and 3-4 bars of pressure are needed. The advancement within the materials requires appropriate stock and process control. The

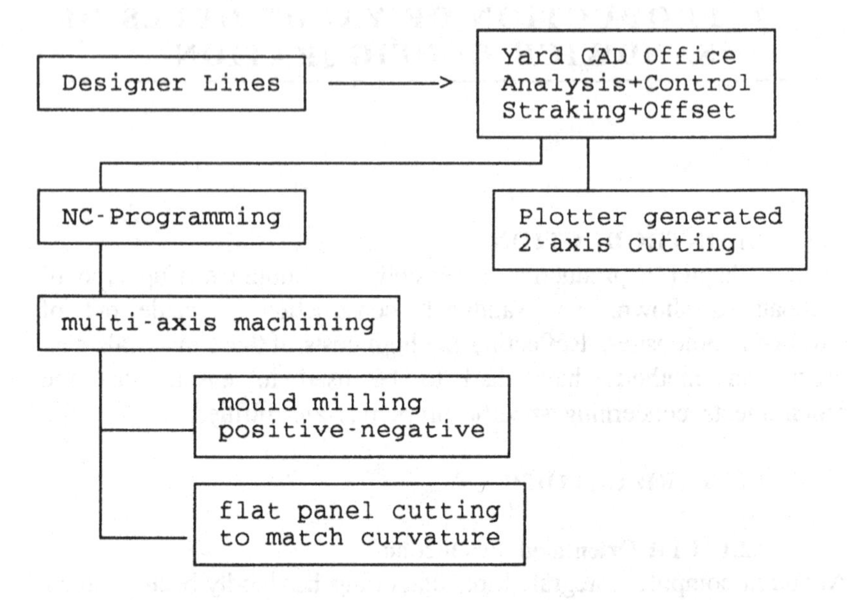

Figure 7.1. CAM concept.

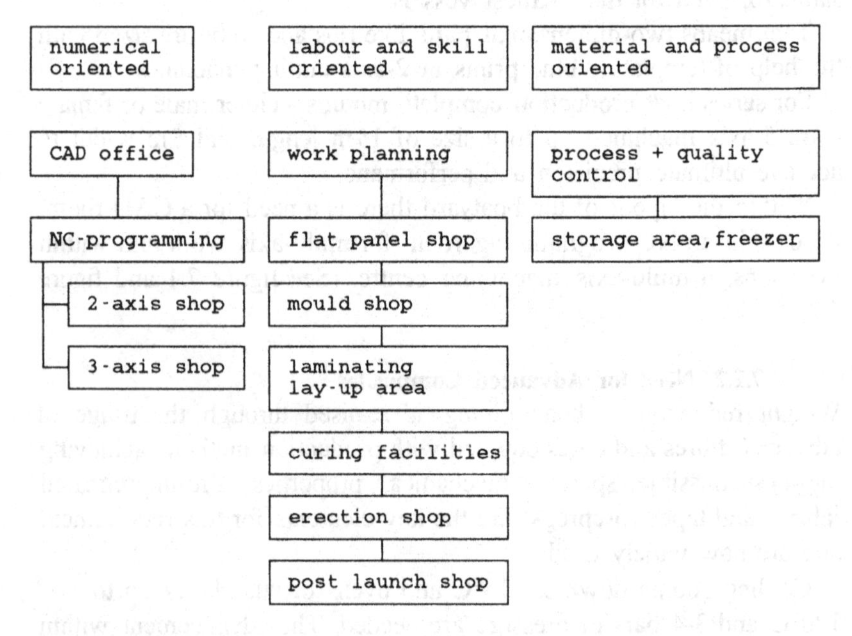

Figure 7.2. Layout requirements.

boatyard layout must have some space reserved for stock and process control.

Prepreg resins carry the curing agents and accelerators already with them. Although the curing agents normally do not react below 90°C there is some polymerisation even at -16°C, which affects tack, drapability and flow characteristics. The flow characteristics are of essential importance when honeycombs are used as core material.

Therefore, at least a climbing drum peel test per resin batch is recommended and repays its price.

7.3 FLAT PANEL SHOP

7.3.1 In-house Production

In most yards electrical, steam or thermal oil heated presses are common and therefore flat sandwich panels can easily be made when temperature control and parallelity of the press openings are given.

Due to the curing temperature of around 120°C required for most epoxy and phenolic prepregs, core materials are limited to the heat resistant PVC foams, PMI foams or honeycombs. The more advanced the fibre lay-up of these panels are, the more honeycombs are used. Here again the flow characteristics are important and "self adhesive prepregs" with flow characteristics as shown in figure 7.3 are needed.

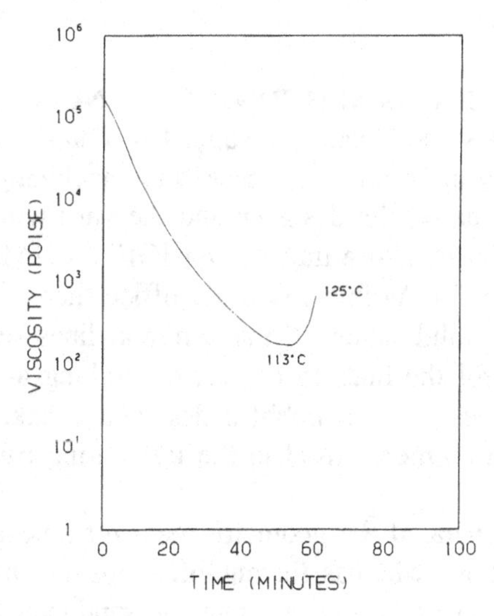

Figure 7.3. Self-adhesive prepreg.

The curing procedure when using epoxy resins is normally hot-in/hot-out, 60 min. at 125°C, pressure 2-3 bars. Advantages of in-house production are the possibility of including hardpoints, inserts or individual reinforcements in one shot and of using advanced prepreg resin technology.

7.3.2 In-house Completion

Sandwich panels are offered today in a wide range of fibre lay-ups and core varieties. For additional laminating work onto the skins, use of peel ply methods can ensure good bonding properties.

Contour cutting is done with the help of the blueprints on a band or circular saw. Tapes are used to achieve good cuts in skins with a high fibre content. Foam sandwiches can be machined to form round edges. Honeycomb panels need edges which are filled with a core filler, normally an epoxy or polyurethane resin filled with microballoons. Loads into the sandwich are transferred with hard points or inserts. Hard points are places where the core has a higher density or where plywood or metal plates are inserted.

It is important that the individual cores are surface treated in the recommended way and that thermal expansion coefficents are matched to achieve long-time durability. In many cases a hole which is reinforced by a collar of core filler is sufficient. For interior fittings and smaller loads pre-manufactured inserts can be used. All inserts and hard points are set in place with cold curing epoxy compounds.

7.4 MULTI-AXIS MACHINING OR NC-SHOP

Advanced materials also require advanced machining capabilities and a smooth way of creating the necessary machining programmes. Normally, the designer and the yard will not have the same software package, so a transfer via IGES, VDAFS, etc. is needed, see figure 7.4. Within the CAD office there should be the ability to create a solid model of the common lines set-up. It is necessary to control the lines in respect of straking and curvature. Special hull features, such as bubbles designed to take advantage of methods and measurements used in the IOR rating rule, will give headaches.

From this is generated the geometric data for a negative mould for series production, whereas for one-off's positive moulds are normally used. These take into account the sandwich thickness and the mould skin and supporting structure thickness (offset).

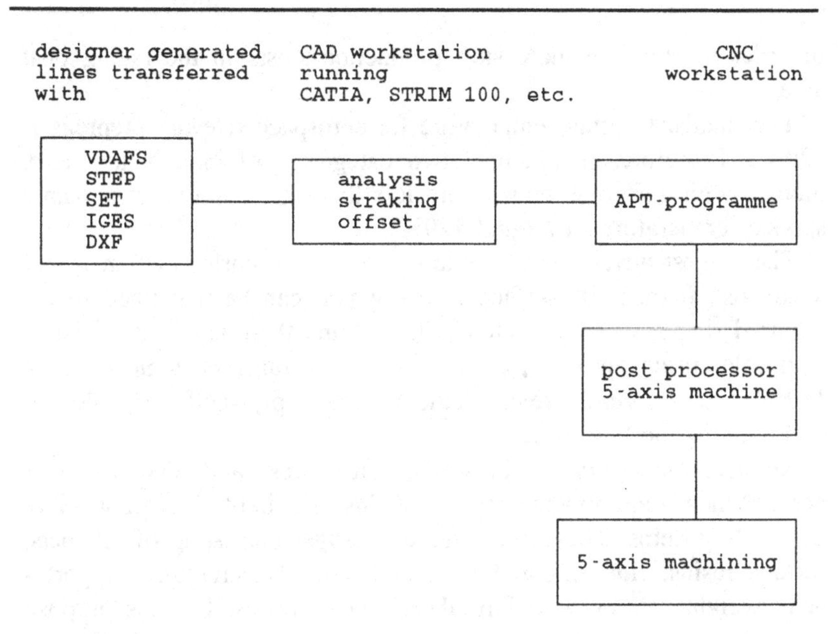

Figure 7.4. CNC concept

5-axis machining today is possible in a nearly unlimited length direction, but height is limited to 2 m and breadth to about 4 m. This means the mould has to be installed on a turntable, where section-wise the contour is machined. The mould for a 15 m yacht, for example, is machined in three sections per side.

Favourable for male moulds are mobile column milling machines, whereas for female moulds the overhead mobile crosspiece milling machine will also work. The gantry type of machine fits for keels and rudders and should be capable of machining metals.

7.5 MOULD SHOP

7.5.1 Materials

Moulds need to withstand the temperatures arising during the curing process, their thermal expansion should be inline with the part being built. The mould should not bend or twist at high temperature. It should be vacuum-tight and its thermal mass should be evenly distributed and not too big. Moulds should be precise and with a super surface. They should be inexpensive.

New production rules restrict allowable curing temperatures, like the IMS (International Measurement System) requirement for 60°C and the America's Cup Class Rule requirement for 95°C. That is done in order to stop the increasing production costs in the racing boat area.

The standard curing temperature for aerospace relevant prepregs is 120°C. Therefore, in practice, two categories of moulds are used, moulds with a service temperature less than 95°C and moulds with a service temperature of around 120°C.

The biggest advantage of the low temperature curing system is that it can use foam built surface layers which can be machined to the required shape, whereas with 120°C systems there is no other choice than cold-curing epoxy systems with a glass transition temperature > 110°C, glass fibre reinforcements and plywood capable of withstanding boiling water.

Rudders are very sophisticated structures and due to their performance requirements, their moulds are built of carbon fibre materials systems. These are material packages consisting of gel coats, surface resins, fine carbon fabrics and - for the structural support - heavyweight fabrics with formulated epoxy resins. For this purpose cold-curing systems with a temperature tolerance up to 130°C are recommended.

7.5.2 Mould Fabrication

- 120°C mould

The author's experience with 120°C epoxy prepregs is that a surface made of two layers of 4 mm cross-plied plywood, bonded and screwed together, covered with one layer of a 120 US style glass cloth, wet laminated, will be in order.

This surface structure is bonded and screwed to longitudinal stiffeners, made out of an easy shapable, light wood. The distance between the stiffeners should be approximately 30 cm and each stiffener should be around 15 x 40 mm, depending on the curvature of the design.

This structure is mounted on 20 mm plywood rib plates, which have an offset outline generated by the CAD solid model. This stands on a sturdy base structure, floating freely on the steel oven cart or trolley.

- 95°C mould

The low temperature mould can be built in the following, advanced way.

Transverse sections each of approximately 0.4 m or 0.5 m are sized by a 5-axis milling machine in such a way that the outline has the

material name	angle of orientation °	compression strength core N/mm²	shear strength material N/mm²	flexural strength N/mm²	flexural modulus N/mm²	tensile strength N/mm²	tensile modulus N/mm²	compression strength N/mm²	compression modulus N/mm²	thickness mm	weight g/m²
Cormaster C1·3,2·48 C1·4,8·64 C1·4,8·96	0 90 0 90 0 90	 1,90 3,20 4,50	0,60 1,24 0,86 1,72 1,15 2,50							40,0 or 30,0	
CF tape 130 g/m² + 90 g/m² EP resin	0 90			1270 650	112000 7000	1250 60	125000 7000	950 200	115000 3000	0,125	220
glass prepreg 107 g/m² + 100 g/m² EP resin	0 90			350 350	16000 1600	270 270	21000 21000	270 270	23000 23000	0,10	214
CF 45° fabric 300 g/m² EP resin wet laminated	0 90			600 600	50000 50000	550 550	60000 60000	450 450	40000 40000	0,40	700
CF tape 300 g/m² EP resin wet laminated	0 90			950 40	80000 3300	900 40	90000 3300	700 200	80000 2500	0,40	600
CF prepreg 370 g/m² + 250 g/m² EP resin	0 90			600 600	40000 40000	450 450	50000 50000	450 420	50000 50000	0,50	620

Figure 7.5. Mechanical Properties of the sandwich construction of CONTAINER '91.

structure of laminate	CONTAINER '83 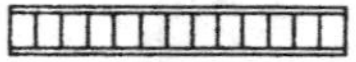 Aramid fabric EP-prepreg	CONTAINER '86 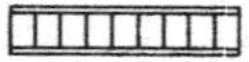 CF tape EP-prepreg	CONTAINER '88  CF tape EP-prepreg	CONTAINER '89, '91  CF tape EP-prepreg	America's Cupper 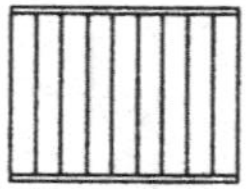 CF tape EP-prepreg
panel weight kg/m²	3,8	3,7	3,6	4,7	11,0
thickness sandwich mm	21,9	21,3	31,0	41,2	54,4
thickness honeycomb mm	20,0	20,0	30,0	40,0	51,0
moment of inertia of outer skin N/mm²	5300	3520	5910	12561	52300
E-modulus*) of outer skin N/mm²	23700	52200	63400	45000	45000
stiffness E I (Nmm²/25,4 mm) · 10^6	125,6	183,7	375,0	565,0	2353

*) E-modulus: average worked out from pressure and tension modulus

Figure 7.6. Sandwich development.

necessary curvature. On this, easy bendable longitudinal sections are bonded and screwed, being 3-4 cm apart. That grid is now closed with a glass cloth (matt) epoxy laminate and a foam layer is placed on top of this. Depending on the required foam skin thickness, two or more layers can be laid-up. Out of this rough mould the 5-axis column milling machine will now precisely cut out the required shape, including any complicated measurement bubbles.

The deck outline on the hull is bounded by a square steel tube, which also acts as a vacuum manifold.

One tempering process before final measurement and production start is recommended. Temperature should be in the range of the following cure, this to ensure that the geometry will respect drawings and drive out entrained gases.

The rudder moulds have to be built precisely according to the data sheet. It is important to work wet-in-wet to avoid gas bubbles and to compress the lay-up by means of vacuum.

7.6 LAMINATING SHOP

7.6.1 Materials

7.6.1.1 Structural Materials

The required materials' lay-up is available from ABS calculations. The production engineer and stress designer should keep in close contact in order not to increase the number of different materials to be used. Figure 7.5 shows the materials and their mechanical properties used for CONTAINER '91, a 50 footer. Figure 7.6 explains how development has moved from Aramid and relatively thin sandwiches to monocoque types of hull with few stiffeners and extremely thin carbon fibre skins. Figure 7.7 explains the normal lay-up procedure. Behind this stands a simple philosophy.

Why choose Nomex honeycombs? This is because weightwise it has the best strength properties, see figure 7.8. Nomex, like foams, is delivered in various densities and individual thicknesses. Figure 7.9 determines what honeycomb density has to be used in the different hull and deck areas. To achieve flexibility honeycombs can be delivered in overexpanded quality to avoid the "saddle effect". Foams for this purpose are delivered in a mat of little foam cubes bonded to a glass fabric substrate.

Why choose an adhesive film? This is because bonding skin to core

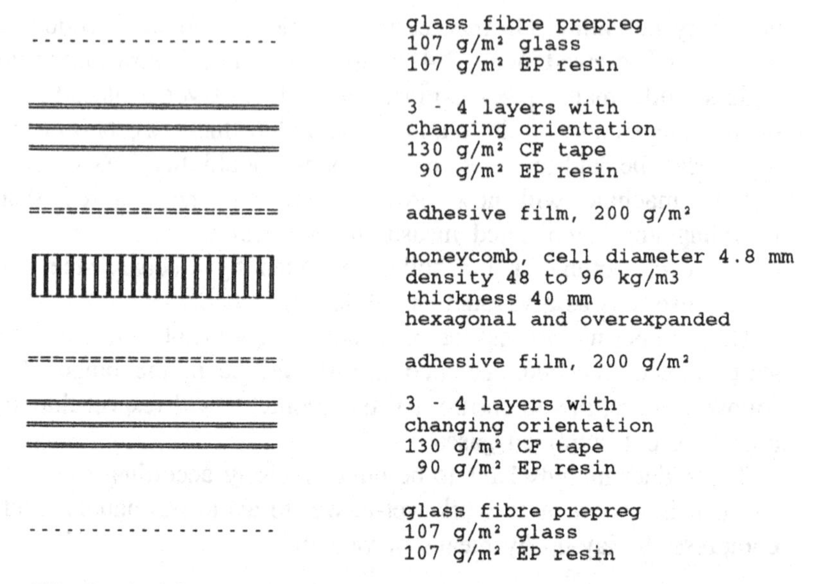

Figure 7.7. Standard lay-up.

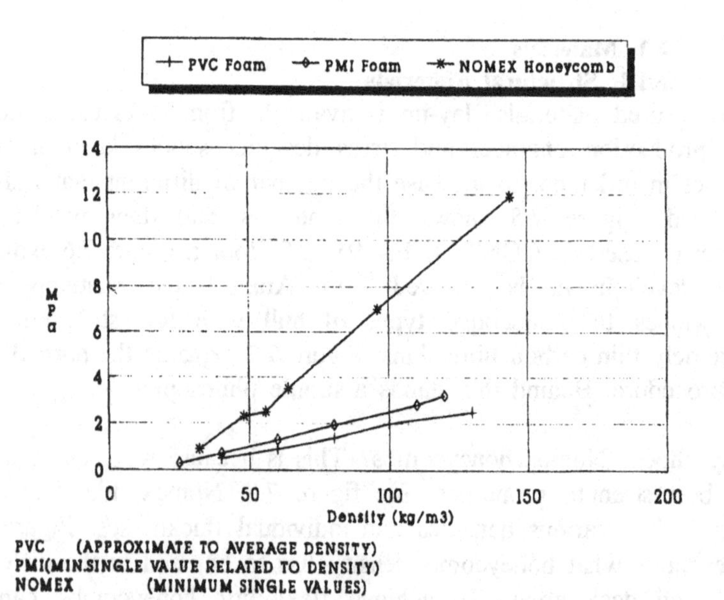

Figure 7.8. Compression strength of core materials.

is the most critical point in prepreg technology. Normal carbon fibre prepreg resins are designed to achieve the best possible mechanical values as composites, they have a high modulus but they are also brittle and poor in elongation and - most importantly - they will not at all bond a prepreg skin to a core, neither foam nor honeycomb. It is necessary, therefore, to use an adhesive film to achieve the right

Panel area	Density [kg/m^3]
foreship deck and above waterline	48
foreship, bow + flat underwater ship	64
midship, cabinhouse top + deck	48
midship, fender area, underwatership	64
keel	96
cockpit, winches	64
stern	48

Figure 7.9. Honeycomb density versus boat area.

fillet bond and a toughened epoxy resin for the wetting of the carbon fibre. Figure 7.10 explains the right bonding behaviour.

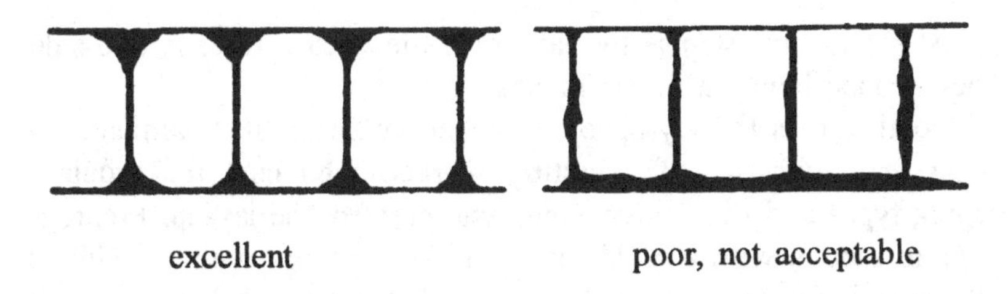

Figure 7.10. Prepreg bonding behaviour.

In line with the toughening of an epoxy resin normally the flow will also be reduced. This is important during the long heat-up times needed for the mould. Heat-up rates between 0.5-1°C per minute are typical, so that the resins are in a very liquid phase for a long time. To avoid dry laminates and poor performance, pre-productions tests to simulate the resin flow to be expected during curing, have to be run to screen the possible resin systems. This procedure is also valid for the outer glass-fabric, which acts as a cover for the carbon fibre structure and also as the finishing substructure.

7.6.1.2 Bagging materials

Male hull moulds are not normally release wax treated, they are covered with used vacuum foil. The vacuum foil should resist the required curing temperature and should be purpose made to avoid a time consuming search for vacuum leaks, or - worst case of all - the loss of vacuum during curing (which would mean a loss of material and labour worth about US$ 50,000-100,000). Vacuum tape is used to seal one vacuum foil to the other and to seal the whole against the mould.

Release agent treated glass or polyamide fabrics, which come on top of the laminate lay-up, do not create just a ready-to-bond surface, but also let the surplus resin flow into the next layer, which is a bleeder felt that stores the surplus resin.

On top of this, a vacuum transfer cloth is needed, which allows overall access for the vacuum. The last layer is the vacuum foil itself.

7.6.2 Production

The lay-up plan is generated out of the ABS calculation schematics. Prepreg materials are taken out of the deep freeze storage into room temperature in sealed bags to avoid humidity condensation. Relative humidity should be around 40% and temperature around 22°C. It is recommended that Nomex honeycomb panels be dried for an hour at 110°C.

First production step is the lay-up of the release film. A two-side adhesive tape bonds it to the mould.

Second step is the lay-up of the inner sandwich skin laminate. A lay-up team consists of a cutting worker, who cuts the required prepreg types and two lay-up men, who perform the lay-up. Prepregs either in fabrics with a width of about 1 m or tapes with a width of 0.305 m will follow their "natural curve" during their lay-up. Much care has to be given to avoid included air bubbles and wrinkles; overlapping is not allowed and gaps are tolerated only if smaller than 1 mm.

The third step is the application of the honeycomb or foam core. Foam has to be needled and a second adhesive layer will be added to ensure bonding. Foam is ideal in shaped parts where Nomex honeycombs can not be formed. Nomex panels have to be laid down according to ABS L and W- designated directions; they can be easily shaped to size with a simple "bread knife".

For ultra lightweight applications it is necessary to learn to bend Nomex core with a heat gun over the hull curvature itself or to prepare simple bending tools. Good examples are the positive and negative radii of the cockpit, which work easily with overexpanded core. Nomex needs a forming temperature of at least 250°C to achieve a standing curvature.

Honeycomb panels are bonded together with a foam tape adhesive, which will expand during cure to 2-3 times its normal volume. The whole honeycomb set-up is fixed to the mould through a net of polyamide ropes, which is attached to the vacuum tube.

It is recommended to bond the whole deck outline to the vacuum tube or, at least, to apply every meter a patch 0.1 m long. This is important because after cure there is only a one skin-sided sandwich consisting of a layer of approximately 0.5 mm CFRP and a core of 30-40 mm. The core will still try to get back to its original form. It bends, twists and moves the sandwich from the mould, creating trapped gas volumes, wrinkles and possibly breaking the skin.

The fourth step is the vacuum bagging procedure. A release fabric is laid on top of the honeycomb and overall a rough glass-fabric provides breathing space. The vacuum foil is prefabricated and bonded to the mould with the sealant tape. To release tensions in the foil it is important that at about every meter, a fold of approximately 0.2 m is made. Then the vacuum pump is started and the whole lay-up has to be made leak-tight at a pressure not higher than 0.15 bar. This pressure is essential to obtain good laminates and core bonding. Of great help for the leak testing are quiet surroundings and an ultrasonic leak tester.

Additional to the vacuum tube line, at approximately every 10 m, a vacuum connection should be made, the same should be done every 1.5 m along the edges within the bow area.

The fifth step is the inner side curing. Temperature indicators are fixed to the outer and inner side of the lay-up. Normally, the bow mould temperature will be the coldest and would determine the curing time needed (90 minutes at 120°C).

Temperatures and pressures are monitored to ensure process parameters are according specification. Total curing cycles of around 12 hours are normal and would be carried out during a night shift.

The sixth step is the lay-up of the outer side. After dismantling the bagging liners the bonding of the core to the inner skin and the splice joints between core sheets are inspected. Any defects can be repaired with cold curing techniques. Also, all the chamfers needed to go into the full laminate or the hardpoints are now machined and set into place. The chamfering ratio should be greater than 5.

Thick full laminate construction, for example on the keel area, is now laminated by hand using relatively heavy roving fabrics and a cold-curing epoxy resin, with a glass transition temperature greater than 110°C. This has to be done stepwise to avoid exothermic reaction. Its surface needs regrinding after curing to achieve curvature and the necessary surface activity for further bonding. After vacuum cleaning of the open structure, the outer lay-up according to figure 7.7, is laid inclusive of the bagging layers. In particular, the second curing needs excellent breathing and vacuum.

The deck is made by practically the same procedure. Care has to be taken in concave areas such that no prepreg bridges, which create non-conforming cross-sections or a lack of bonding between core and skins are built.

It is recommended to cut the fabric or tape and work with a suitable overlay. For achieving the necessary vacuum pressure in these radii, pressure bars, shaped to size, are used.

Within the second deck lay-up a lot of hard points have to be positioned. Whether cold bonded or co-cured with the second curing, the surface preparation for bonding has to be done. A cold-curing epoxy putty with glass or plastic micro-balloons is used as a standard bedding compound.

7.7 CURING REQUIREMENTS

Hull moulds are great heat sinks whereas the core or the laminate lay-up can be quickly heated up. Mould temperature and laminate temperature should grow together. To assist normal heat flow large pipes have to be laid into the mould inner side to keep both mould and laminate temperatures close to each other. Therefore, heat-up ratios greater than 1°C/min. for the circulating air should not be used. The ultimate temperature should be 10°C higher than the curing temperature, to avoid long heating time and increased reactivity. For easy access and air circulation at least 0.5 m width at both sides should be added to the mould size. Above the mould, room should be allowed for a central vacuum tube which transfers the vacuum suction through individual plastic tubes and connections into the laminate lay-up.

For safety reasons two vacuum pumps in a parallel mode with a power of about 1.5 kW should be used for the evacuation and should be run continuously. Temperature variation within the mould surrounding should be limited to around 5°C.

Ovens need to be well isolated. Heat generators should work as heat exchangers and not as direct burners. This is to keep the oven clean and - more importantly - to allow access for inspection during cure.

Rudders and masts in these days are such extreme constructions that autoclave curing is required. Autoclaves for these applications should have a capacity for pressures up to 5 bar and curing temperatures up to 170°C. Rudders can be built in the aerospace or racing car type of autoclave of 3-4 m in diameter, 6 m in length, whereas masts and spars require long tunnel autoclaves that arc available only at specialist manufacturers.

7.8 ERECTION SHOP

One big disadvantage of working with a male hull mould is the need for extensive finishing work on the outer hull surface. Although prepreg laminate surface imperfections are normally in the range of 0.15-0.4 mm, vacuum forces will create wrinkles, which have to be filled with special lightweight epoxy fillers. Filling, curing and grinding to a smooth surface is time and cost-intensive. A lot of craftsmanship is needed to achieve a good surface quality without adding too much weight to the base structure. The finishing work is done when the hull is still on the mould to avoid overhead working.

The hull is then turned into its normal position and lowered into a number of saddles. Sandwich hulls are self-supporting monocoque structures but without a deck they are flexible. Therefore, a temporary frame is needed to set the hull. In the keel area a longitudinal berth support is needed to carry the loads of the inner ballast.

In the early eighties the keel was attached to an aluminium frame, which was set into the hull with lots of epoxy putty. The production of this frame was very expensive and its function questionable. Leaking keel bolts and a separation of the hull from the frame were normal. A big improvement is the technique whereby the inner lead ballast is designed as a load carrying unit that the keel is attached to, see figure 7.11.

Lead bars are cast with a trapezoidal cross-section of the curvature of the inner hull. Metal thread devices are drilled into the bars, to ease handling in setting them into their epoxy putty bed. A stainless steel plate on top of this will give the optimal keel bolt connection and guarantee structural safety under racing conditions.

The mounting frames for the engine, the hydraulic pump, etc. - normally made out of wood - are then laminated into the hull. Also, all bulkheads, webframes and longitudinal stiffeners are attached with

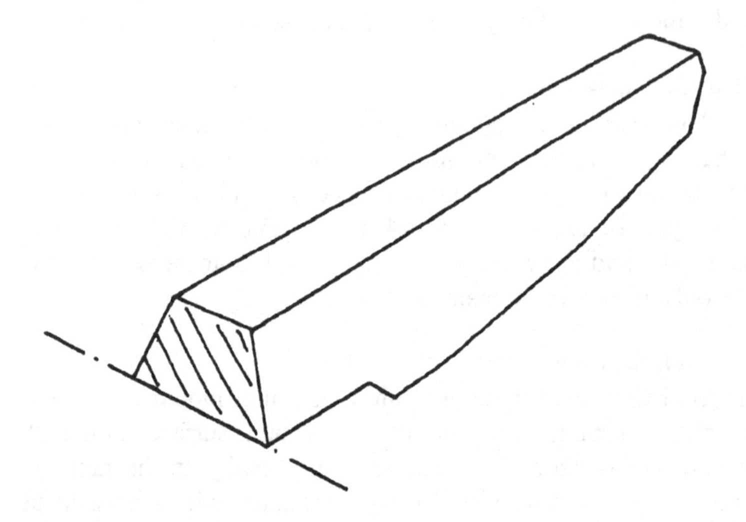

Figure 7.11. Inner lead ballast.

epoxy putty to the hull and then connected with one layer of a ± 45° carbon fibre tape.

The mast is stepped on a wooden bearer between the main mast bulkhead and the auxiliary half bulkhead. The main forces from the mast are transferred via a heavy glass laminate into the bulkheads.

In the slam prone areas especially, spacing between load carrying members should not be more than 1.5 m. Special care is also needed in highly loaded areas like the shroud connections - here again the load is transferred into the two mast bulkheads via local stiffening.

In the same way as the hull, the deck is reinforced with stiffeners when still lying in its mould. The cabin hatchway is cut out and the surplus part used for quality purposes. The hull is covered with epoxy putty where it joins the deck and the deck correctly set in place and attached.

Again ± 45° carbon fibre tapes are used to connect the hull to the deck and the deck structure to the hull stiffeners. The last structural point is the integration of the rudder bearings and rudder tube into the cockpit and hull structure.

Modern rudders produce large forces and moments, so two sandwich panels are used to transfer loads from the tube into the cockpit after bulkhead.

The openings for the shrouds and masts have to be cut in the deck, the guard rail fittings have to be set, cabin hatch and other items have to be laminated, filled and ground before the boat goes to the paint shop.

7.9 POST LAUNCH OUTFIT

An Admiral's Cup boat is built to win races; so the outfit items need to enhance operational efficiency.

The inner outfit is directed towards getting sails changed quickly and stored under deck. Sleeping, toilet and cooking facilities are lightweight and simple. Navigation aids and performance monitoring aids will be integrated in the navigators place. On deck the fitting of winches, sheeting blocks and other gear requires a precise installation each item in its proper place, where the appropriate reinforcements have been fitted.

7.10 QUALITY ASSURANCE

Racing boats are built for ultimate performance, consequently safety factors are reduced to a minimum. Sailed under racing conditions excessive strain may occur; therefore a lot of care and supervision is needed to ensure a good, safe boat.

The laminate quality is tested with long beam bending specimens and climbing drum peel tests to confirm the structural performance and integrity. Delaminations are checked using the simple "coin" method. Accurate 'as built' dimensions are recorded by the surveyor in the Certificate of Measurement.

8 MATERIAL CASE STUDY FAILURES AND THEIR REPAIRS

8.1 INTRODUCTION

This Chapter is intended to give an insight into the practical considerations associated with the failure of FRP materials and/or structural components. From this, it will be appreciated that one of the major considerations is the definition of "failure" and, consequently, this can differ depending upon the intended use of the component. From the marine industries' point of view "failure" of a laminate must be the initial surface fracture that would permit moisture penetration which, if not satisfactorily dealt with, will lead to subsequent deterioration of the laminate.

From the above, it will be appreciated that identifying the individual types of failure is only one of the practical considerations and unless accompanied by adequate repair procedures, the structural integrity of a component cannot be reinstated.

Prevention of "failure" can generally be taken into consideration in the design and fabrication process. Therefore, details of precaution against failure for laminates and hull structures have been included in this Chapter.

8.2 GEL COAT FAILURE

The gel coat should be of a uniform thickness throughout the surface of the mould. A gel coat of uneven thicknesses, or where the catalyst has been poorly mixed, will cure at different rates over its surface causing stresses which could lead to crazing and also give a patchy appearance. A very thin gel coat may not cure correctly and can result in attack when the resin for the back-up reinforcement is applied, leading to the possibility of gel coat wrinkling. Thixotropic gel coats minimise this occurrence.

A Barcol hardness indicator should be used to ensure that the gel coat resin has correctly cured after demoulding. The best way of examining the gel coat is visually. The following section gives examples of common faults found on the outer surface to the gel coat.

8.2.1 Common Faults

Wrinkling: This is caused by solvent attack on the gel coat by the monomer in the laminating resin and, in general, is due to the fact that the gel coat is undercured. Wrinkling can be avoided by ensuring that:-

- the resin formulation is correct
- the gel coat is not too thin
- the catalyst level has not been reduced below 1%
- the temperature and humidity are controlled
- air is not blown across the exposed surface (especially if it is warm air) which would lead to a reduction in monomer level and hence a faulty cure

Pinholing: Surface pinholing is caused by small air bubbles which are trapped in the gel coat before gelation. It occurs when the resin is too viscous, or has a high filler content, or when the gel coat resin reacts with the release agent. Care must also be taken to ensure that the mould surface is clean and free of dust.

Poor adhesion of the gel coat resin: Unless the adhesion of the gel coat to the backing laminate is very poor, this defect will be noticed only when the structure is being handled and pieces of gel coat flake off. Areas of poor adhesion can be detected sometimes by the presence of a blister, or by local undulations in the surface when it is viewed obliquely. Poor gel coat adhesion can be caused by an inadequate quantity of resin between the gel coat and back-up reinforcement, poor consolidation of the laminate, contamination of the gel coat before the glass fibre is laid-up or, more generally, by the gel coat being overcured.

Spotting: This fault takes the form of spots all over the gel coat surface of the laminate. It is usually due to one of the ingredients of the resin formulation not being properly dispersed.

Striations: This fault is caused by pigment flotation and is most likely where the colour used is a mixture of more than one pigment. The remedy is thorough mixing or the use of a different pigment paste.

Fibre Pattern: The pattern of the glass fibre reinforcement is sometimes visible through the gel coat or prominently noticeable on its surface. This usually occurs when the gel coat is too thin or when the reinforcement has been laid-up and rolled before the gel coat has hardened sufficiently, or when the moulding is removed too soon from the mould.

'Fish eyes': On a very highly polished mould, particularly when silicone-modified waxes are used, the gel coat sometimes 'de-wets' from certain areas leaving spots where the gel coat is almost non-existent. This shows up as patches of pale colour usually up to 6 mm in diameter. It can also occur in long straight lines following the strokes of the brush during application. This fault is rarely experienced when a PVA (Polyvinylalcohol) film is correctly applied.

Crazing: Crazing can occur immediately after manufacture or it may take some months to develop. It appears as fine hair cracks in the surface of the resin. Often the only initial evidence of crazing is that the resin has lost its surface gloss. Crazing is generally associated with resin rich areas and is caused by the use of an unsuitable resin or resin formulation in the gel coat. The addition of extra styrene to the gel coat resin is a common cause. Alternatively, the gel coat resin may be too hard with respect to its thickness. In other words, the thicker the gel coat the more resilient the resin needs to be. Crazing which appears after some months of exposure to the weather or chemical attack is caused by either undercure, the use of too much filler, or the use of a resin which has been made too flexible.

Star Cracking: This is the result of having an over thick gel coat and occurs when the laminate has received a reverse impact. Gel coats should in general, not be more than 0.5 mm thick, unless the gel coat is rigidly supported. Regarding this type of defect, it should be remembered that impact damage on the mould surface will be reproduced in the surface of the moulding. Therefore, particular attention should be given to checking this type of defect before implementing any repair.

Internal Dry Patches: These can be caused by attempting to impregnate more than one layer of reinforcement at a time and also where heavyweight reinforcements are used (i.e. greater than 1200 g/m^2). The presence of internal dry patches can be readily confirmed by tapping the surface with a coin.

Poor Wetting of the Mat: The cause of poor wetting of the mat is either the use of insufficient resin during lay-up, or inadequate consolidation of the lay-up. This defect is normally apparent on the reverse face of the laminate only, i.e. the side without a gel coat. When correctly wetted this will have a glazed appearance because the fibres are coated with resin.

Leaching: This is a serious fault. Leaching occurs after exposure of the laminate to the weather, and is characterised by a loss of resin from the laminate leaving the glass fibre exposed to attack by moisture. Leaching indicates either that the resin used has not been adequately cured, or that it is an unsatisfactory resin for that particular application.

Yellowing: GRP laminates yellow after a period of exposure to sunlight. It is generally only slight, but can be considerable on translucent roof sheeting and white pigmented laminates. It is a surface phenomenon due to the absorption of ultra-violet radiation. For this reason, most sheeting resins contain UV stabilisers which reduce considerably the rate of yellowing. Yellowing has little effect on the mechanical properties of the laminate.

Typical examples of these defects are given in figures 8.1-8.13.

8.2.2 Gel Coat Back-up

To provide an adequate water barrier for the laminate, it is vitally important that a resin rich powder bound CSM back-up reinforcement is applied to the gel coat and, in general, this should have a maximum weight of 300 g/m^2. Past service experience has shown that emulsion bound mats should be avoided and that the provision of a good primary bond to the gel coat is essential.

However, the application of this CSM back-up gives rise to two conflicting considerations:-

i) It provides the required water barrier, and
ii) It places the weakest reinforcement at the maximum distance from the neutral axis of the laminate.

Hence, due to local deflection of shell panels, the CSM is subjected to the highest loads. Consequently, failure of the CSM tends to occur first and may lead to a chain reaction of degradation. In this respect, the CSM controls the load carrying capabilities of the laminate and

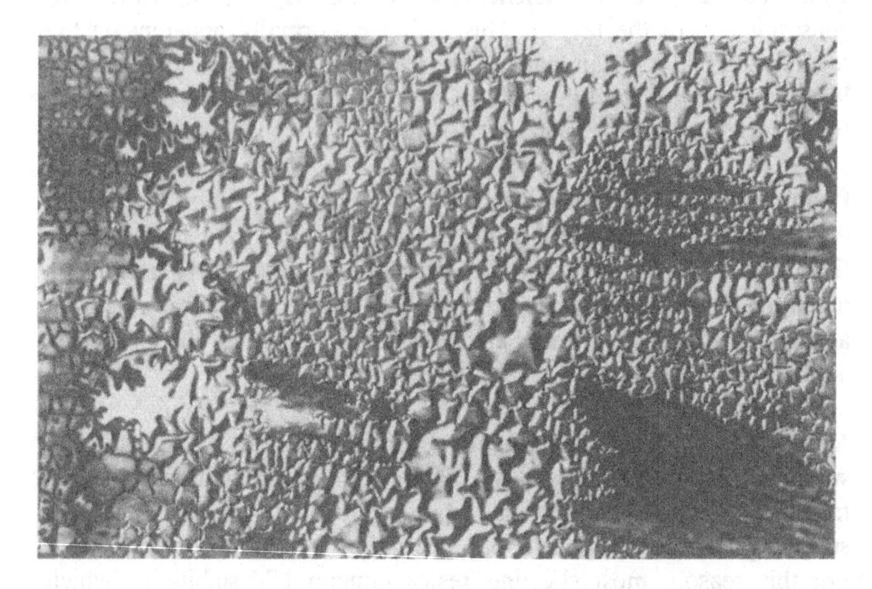

Figure 8.1. Wrinkling.

Figure 8.2. Pinholing (photomicrograph).

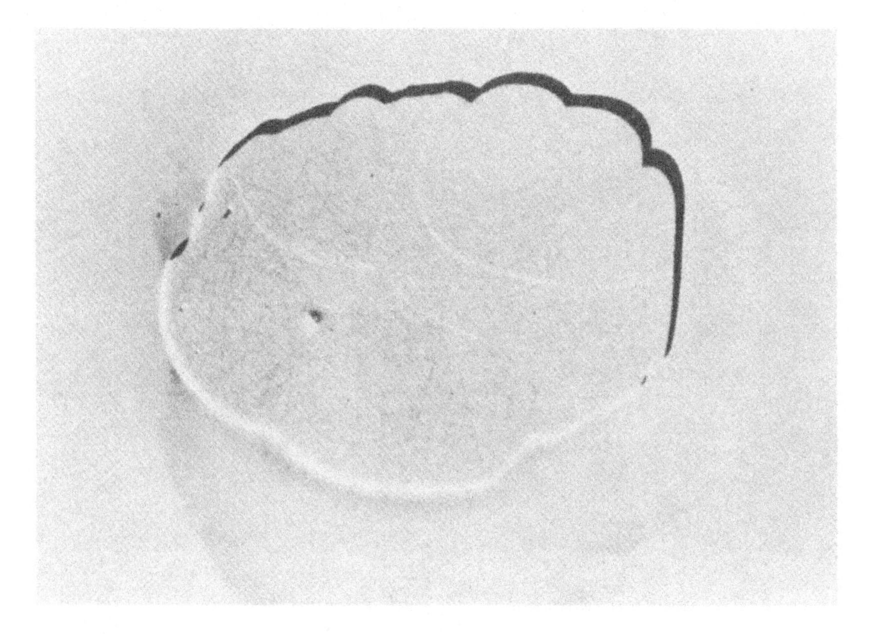

Figure 8.3. Poor gel coat adhesion.

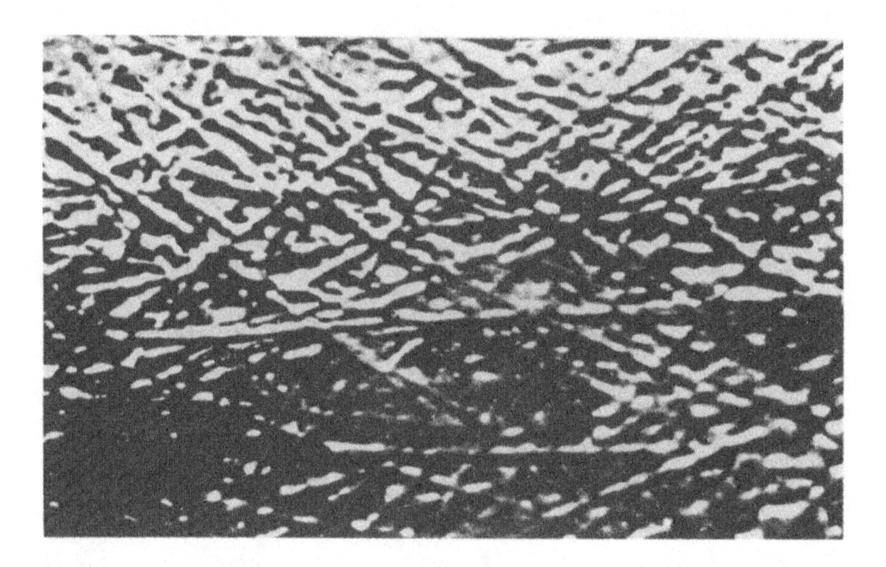

Figure 8.4. Fibre pattern.

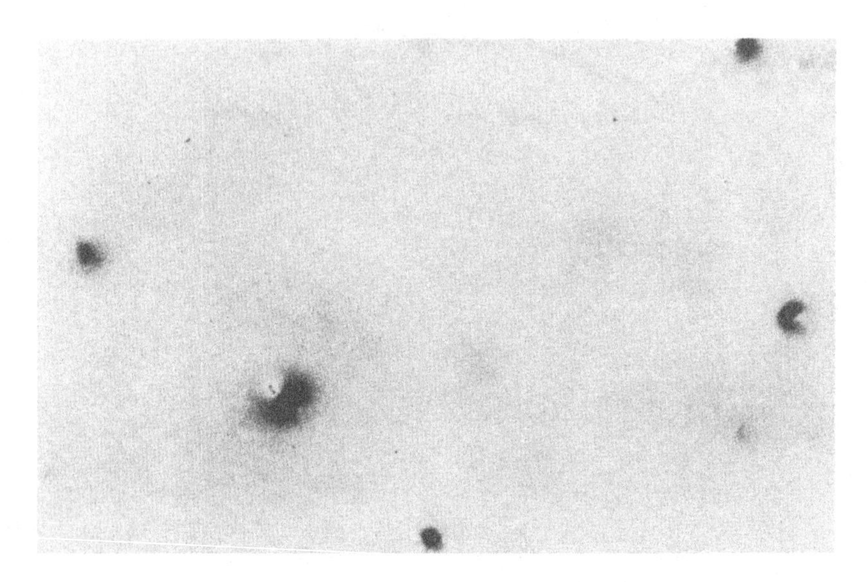

Figure 8.5. 'Fish eyes'.

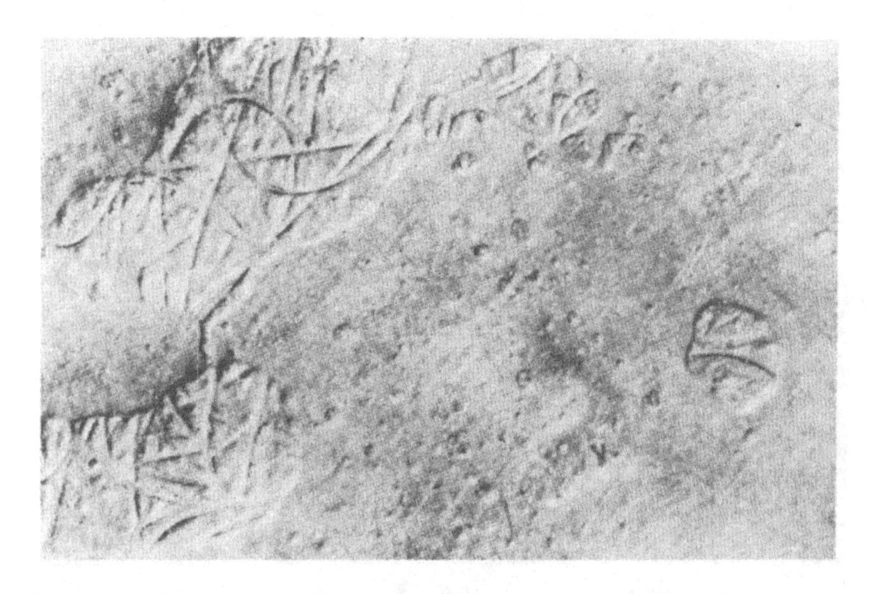

Figure 8.6. Severe internal blisters seen through translucent gel coat.

Figure 8.7. Crazing (photomicrograph).

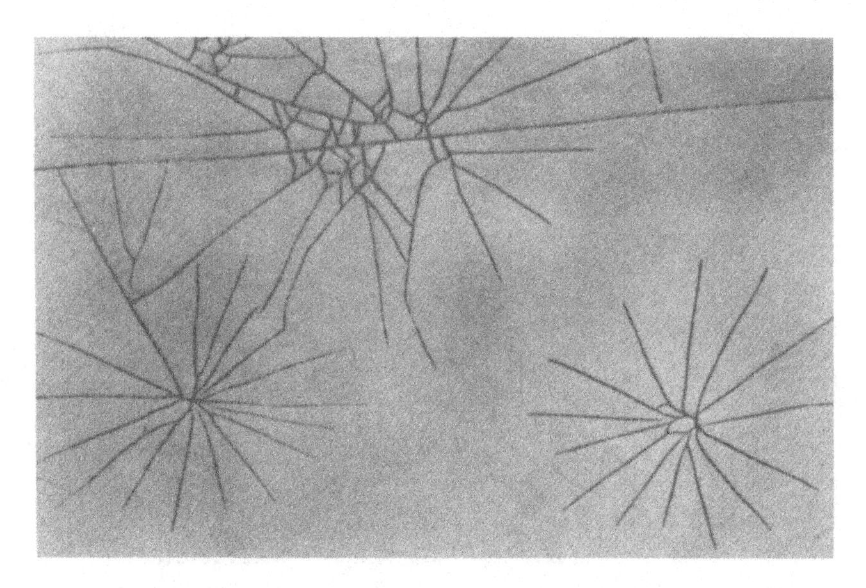

Figure 8.8. Star cracking.

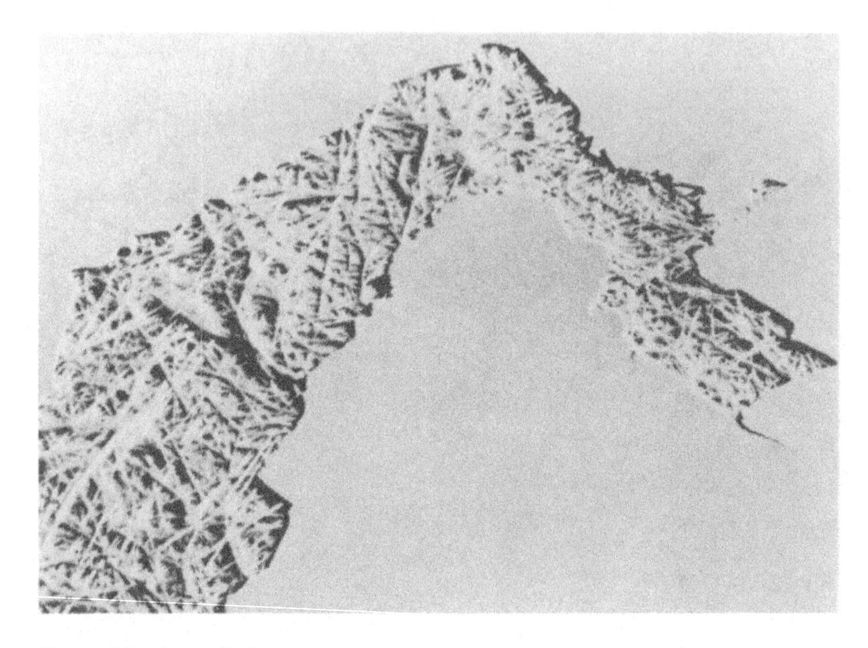

Figure 8.9. Internal dry path.

Figure 8.10. Severe leaching.

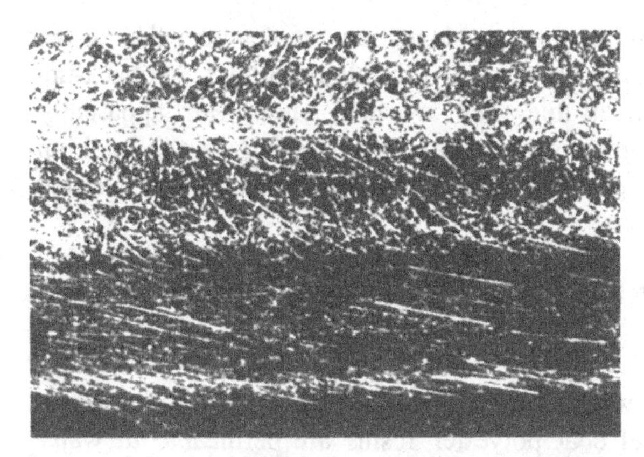

Figure 8.11. Faults associated with translucent sheeting - fibre pattern.

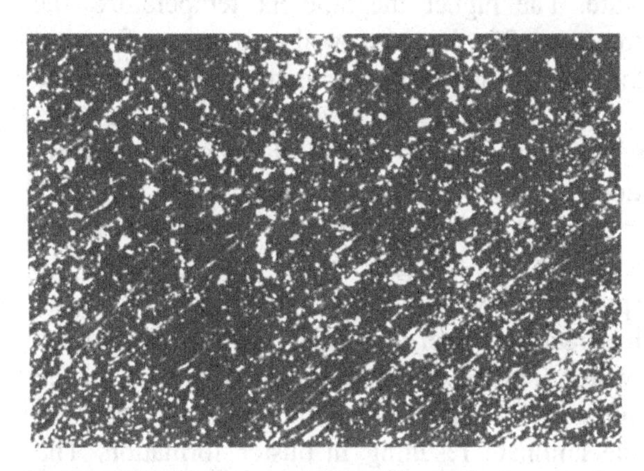

Figure 8.12. Faults associated with translucent sheeting - speckling (undissolved binder).

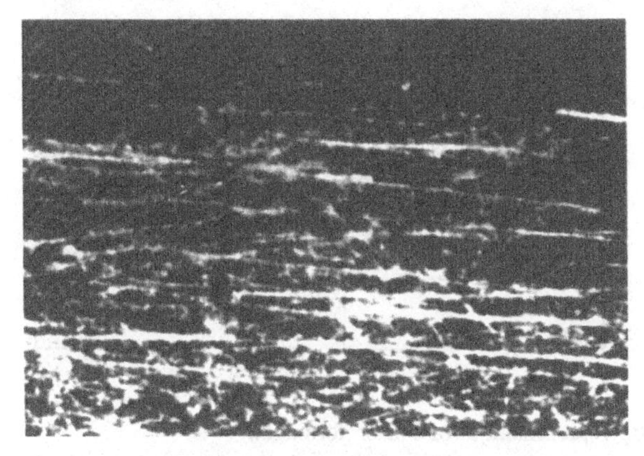

Figure 8.13. Faults associated with translucent sheeting - flecking.

failure of gel coat, (which has an elongation greater than that of the lay-up resin), is generally associated with failure of the CSM back-up. Therefore, it is conceivable that a craft having an apparently undamaged gel coat may suffer from overstressing of the back-up laminate and may be prone to fatigue cracks forming in the gel coat under certain conditions.

8.3 MECHANICS OF GEL COAT BLISTERING

8.3.1 General

GRP is not entirely "waterproof"!

Laminating and gel coat polyester resins are permeable to water vapour and water vapour migration occurs through the laminate at a constant predictable rate. The higher the ambient temperature, the greater the rate of permeation. The laminate will continue to function without problem as long as the water vapour continues to permeate through. If, however, water molecules are actually able to condense in the laminate then this can lead to one of the major problems namely blister formation.

Blistering can occur between the gel coat and the laminate or in the laminate itself. The reason is thought to be the presence of hydrophilic centres (possibly low molecular weight glycols, acid, surfactants etc.) which attract the water molecules and form a solution of the hydrophiles in water. This results in the ideal condition for the formation of the osmotic cell and the resulting pressure leads to delamination within the laminate resulting in blister formation. The process is shown diagrammatically in figure 8.14. Further information is available in reference [1].

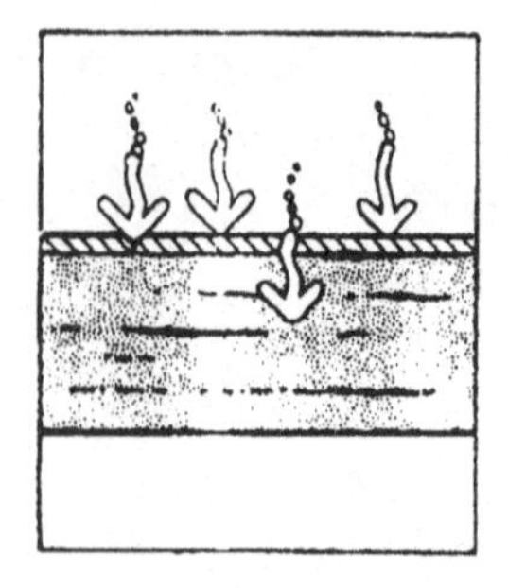

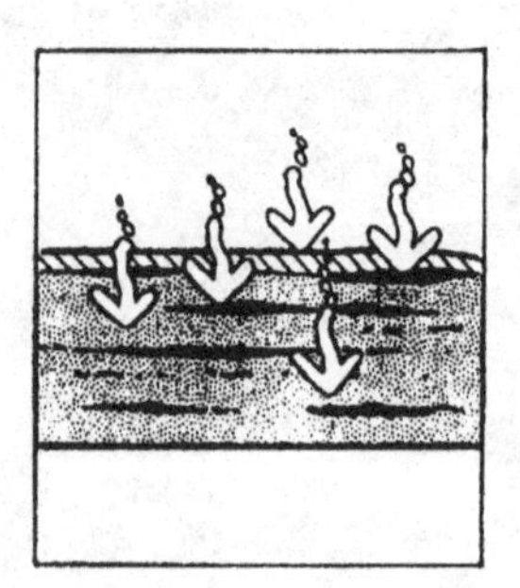

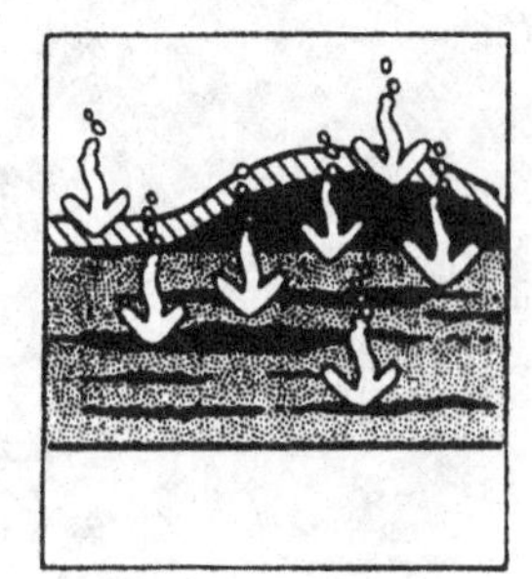

Figure 8.14. Diagrammatic representation of blister formation.

8.3.2 Factors Influencing Blister Formation

The question of the factors effecting blistering is very complex and offers no simple reasons as to the cause. The range of variables is very broad, as the following factors indicate:-

- the polyester type of gel and laminating resins and the selection of the catalyst/accelerator system can result in poorly mixed or incompletely reacted matrices
- additives to the resin system such as thixotropic agents, hydrophilic fillers, pigment pastes, fire retardants and the use of excessive styrene as a diluent, can all change the sensitivity of the moisture penetration rate and also stimulate hydrolysis
- gel coat thickness (either over or under thickness) or of a resin system with poor moisture permeability resistance
- resin to fibre bond, the nature of the finish, and the type and quantity of binder. In general powder bound CSM should be specified as a "back-up" material to the gel coat
- void content (including air bubbles and cracks) which can be effected by both the degree and rate of cure
- fresh water is a higher risk than salt water and also temperate water is worse than cold water
- poor workshop environmental conditions leading to resin undercure
- poor moulding practices resulting in poor adhesion between gel coat and laminate and poor interlaminar bond
- overcure before laying-up the next layer

8.3.3 Methods to Reduce Blistering

Blistering in GRP can be reduced by using:

- only clear or white isophthalic acid based gel coats below the waterline
- a resin rich surface-tissue layer behind a gel coat
- resins which are less susceptible to hydrolysis such as those based on isophthalic acid
- reinforcements containing non-hydrolysable binder, especially with orthophthalic acid based resin system
- fabrication techniques resulting in complete wet-out with minimal void content.

8.3.4 Condensation Testing

Extensive testing was undertaken in the 1980s by DSM Limited using the Amoco Turtle Box and the QCT Condensation Tester Methods. The QCT creates a 100% humidity at temperature settings between 40 and 80°C. In the case of coatings, it has been shown that a condensation tester instead of a humidity tester is more severe in reproducing corrosion effects.

The QCT provides a positive sample cooling by room air so that there is a continuous and predictable supply of freshly condensed water on the sample. The effect of exposure on the gel coat laminates is quite dramatic and at a temperature of 60°C, it has been found that blistering can be reproduced in weeks depending on the type and nature of the composite being tested. Blisters of all shapes and sizes from a surface roughness to centimetres in diameter have been produced. The effect of different permeation rates of water through gel coats and laminate is clearly accelerated using this equipment which provides a simple and quick way of studying improvements in water resistant systems. The condensation tester has been successfully used to compare blister resistance of many systems with consistent results.

Sample construction and preparation conditions are the same, there are many factors which effect blister resistance of gel coated laminates for example, thickness of gel coats or laminate, glass-type and presence or absence of surface veil or pigments. For further details of this particularly relevant system, see reference [2].

For additional information concerning the concept of "total control" of osmotic conditions, see reference [3], where a comprehensive description of various effects of different resin additives is given.

8.4 PRECAUTIONS AGAINST FAILURE IN LAMINATE DESIGN

8.4.1 Lay-up Considerations

A great deal is spoken about the use of advanced composites for sandwich laminate design; however, general practice in yachts and small craft is to use conventional laminates featuring E-glass/polyester resin. Usually, even advanced sandwich laminate hulls are designed with single skin laminates in way of concentrated loads such as keels, chines, masts and chainplate attachments etc.

Whilst the mechanical properties of various composite reinforcements are available from manufacturers, there is unfortunately a limited supply of accurate information for the mechanical properties of very advanced composites. Just how necessary it is to use so many different materials in different forms in laminate construction is not entirely clear. Numerous manufacturer's propose various weave patterns featuring biaxial, triaxial, quadriaxial material which have dubious benefit to the average yacht and boat designer.

A tabulation for typical material properties for laminates is given in table 8.1.

An optimally designed shell or deck laminate would have strength and elastic properties in each in-plane direction to match the loading in that plane and enable each ply in the laminate to be used as efficiently and effectively as possible. By consideration of this standard, there are probably few if any, optimum design laminates available. The problem becomes a trade-off between design effort and cost benefit.

The primary directions of loading and stress on a boat hull or deck panel are longitudinal and transverse. Strength and stiffness of a laminate in these two directions can obviously be used to advantage. Where high torsion or high shear loads exist, plies with substantial fibre reinforcement in the ±45° axis can be used to advantage. Where loads apply primarily in one direction, unidirectional reinforcements can, in general be used to full advantage. Examples are internal structural members such as frames, longitudinals and girders. However, certain of these are also subject to torsion such as rudder stocks which may adopt reinforcement in the ±45° directions. Unidirectional materials may also be laid at 45° in way of the corners of large openings which are subject to high stress concentrations.

In considering the various plies to be used in a laminate, the compatibility for efficient use can be assessed by comparing the strength to modulus of elasticity of the different plies of the laminate. This will allow a study of efficiency of the overall laminate and establish any areas which are not being utilised and other areas where relative overload of a particular ply may take place. This problem is particularly relevant when selecting a laminate thickness where a CSM is used on the outer surface against the gel coat to provide a water barrier. The CSM is relatively weak and may have a strain to failure of 1.3%.

In both single skin and sandwich laminates, there should be a reasonable similarity in the type and sequence of plies used on each side of the neutral axis. Likewise, unless a more detailed analysis is undertaken to assess the effects, mechanical properties of the inner

Table 8.1. Typical material properties of reinforcements.

Material	Matrix	Fibre Weight Fraction	Density (g/cm^3)	Ultimate Tensile Strength (N/mm^2)	Tensile Modulus (kN/mm^2)	Ultimate Compressive Strength (N/mm^2)	Specific Tensile Strength	Specific Tensile Modulus
Higher Tensile Steel	-	-	7.75	480	207	340	62	27
Aluminium Alloy 5083 0 (N8 0)	-	-	2.66	275	69	275*	103	26
Aluminium Alloy 6061 T6 (H30)	Welded	-	2.7	165	69	165*	61	25
	Unwelded	-	2.7	289	69	289*	107	25
E-Glass Random Mat	Polyester	0.33	1.44	94	7.5	122	65	5.2
E-Glass Woven Roving	Polyester	0.50	1.63	188	13.8	147	115	8.5
S-Glass Woven Roving	Polyester	0.50	1.61	400	18	200	248	11
Aramid (Kevlar 49) Woven Roving	Polyester	0.45	1.30	298	21	96	229	16
Carbon Fibre Woven Roving	Polyester	0.40	1.40	450	30	-	321	21
Aramid (Kevlar 49)	Cold cured Epoxy	0.46	1.31	390	21	180	344	16
Carbon Fibre Woven Roving	Cold cured Epoxy	0.54	1.47	500	50	310	340	34

* Approximated at same value as tensile properties

and outer skins of the sandwich panel should be similar in the same laminate in-plane directions. If there is an appreciable difference between the skin properties in the major in-plane axes, the effect of this on load distribution and strength of the laminate should be determined.

When designing sandwich laminates, it must be remembered that provided the core has been designed correctly, failure of the skins will, in general, only occur in way of details and consequently, particular attention should always be given to the design of these areas. A typical problem that causes failure in sandwich construction are where the core materials are used to transmit the loading around changes of section such as chines, etc. These problems can be obviated in design by incorporating single skin areas or shear ties in a sandwich core. Particular attention must also be given to the skin thickness in way of the transitions to ensure that the outer skin is always greater than the inner skin plus any subsequent internal bonding which will be incorporated later.

The direction and sequence of laminating can reduce the extent of failure. This is particularly important on large fast craft where minor impacts can result in considerable damage due to the water pressure delaminating layers of reinforcement or even the whole of the outer skin of sandwich laminates. To obviate these problems, particular attention should be given to the forward end of the craft regarding the following:-

i) Incorporating a sacrificial area in the stem.
ii) Making sure that the lay-up sequence and direction of reinforcement is such as to reduce and not increase the extent of delamination, see figure 8.15.

The above comments also apply to laminates used for internal stiffening members such as frames, longitudinals, beams and girders. However, the emphasis of laminate balance in this case is the balance of the top hat flange formed by the shell or deck in the direction parallel to the internal member. Whilst the design of top hat stiffeners is fairly well documented, very few occasions can be identified where the emphasis is placed on shear considerations, see figure 8.18 and compressive strength calculations. All elements of the stiffener should have adequate stability and where effective cores are fitted they should be checked to ensure that they fulfil their supporting functions.

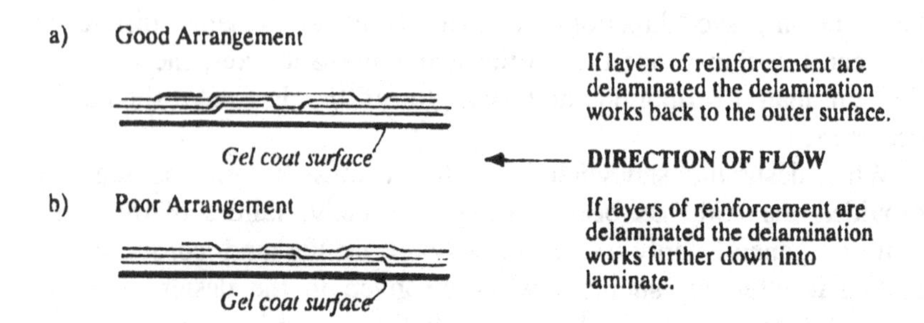

Figure 8.15. Shell lay-up alternatives.

Core materials may be of many varying forms ranging from PU (polyurethane), PE (polyethylene), PVC (polyvinylchloride) and plywood. All of these materials are relatively easy to design and mechanical test data may be used to support the calculations made. Before using core materials of wood and plywood, their strength to modulus properties should be checked against those of the GRP for the frames or other internal stiffening members since their relatively low strength to modulus properties generally indicate tensile or compressive failure of the wood core before the failure of the GRP laminate.

8.4.2 Standard Composites

In general, standard composites are E-glass laminates with alternate plies of CSM and woven rovings. Typical room temperature cure properties for design are given by the following formulae:-

	N/mm^2
Ultimate tensile strength	$127G_c^2 - 510G_c + 123$
Tensile modulus	$(37.0G_c - 4.75)10^3$
Ultimate flexural strength	$502G_c^2 + 106.8$
Flexural modulus	$(33.4G_c^2 + 2.2)10^3$
Ultimate compressive strength	$150G_c + 72$
Compressive modulus	$(40.0G_c - 6.0)10^3$
Ultimate shear strength	$80G_c + 38$
Shear modulus	$(1.7G_c + 2.24)10^3$
Interlaminar shear strength	$22.5 - 17.5G_c$

Where G_c = the glass content of laminate, by weight (excluding the gel coat) determined from the formula:-

$$G_c = \frac{2.56}{\frac{3072\ T}{W} + 1.36} \tag{8.1}$$

T = the nominal laminate thickness, in mm
W = the total weight of glass reinforcement in the laminate, in g/m^2

The properties of these laminates allow relatively simple structural design and construction. The use of sandwich construction or closely spaced internal stiffening with single skin laminate can offset the structural flexibility associated with the tensile, compressive and flexural moduli of standard laminates. There are other woven and non-roving forms of E-glass reinforcement including unidirectionals that are not used as effectively as their potential allows. The unidirectional material allows for very high modulus stiffeners to be produced quickly and cheaply. The only warning that should be made with the use of unidirectional materials is that web thickness of stiffeners must be checked to ensure that web buckling does not occur.

8.4.3 Advanced Composites

Reinforcing fibres for advanced composites are generally other than E-glass (such as S-glass or R-glass) and may include aramid or carbon fibres. Each material has a number of significant and positive characteristics with often one less useful feature. Typically whilst aramid fibres demonstrate very high modulus and light weight characteristics, the compressive properties particularly in the wet conditions are low.

Because of the problems associated with each individual material, they are often combined together to provide a hybrid material which compensates for the deficiencies of one material whilst using the benefits of another material. Typically, Aramat 72K is a good example where the combination of properties has been used to good effect and is widely used in a large number of applications.

Significant improvements in laminate characteristics can also be achieved by selecting the correct weave type and by careful distribution of the various components in their various directions.

The greatly superior properties offered by these hybrid materials can vary tremendously with laminate fibre direction and typical characteristics are shown in figure 8.16.

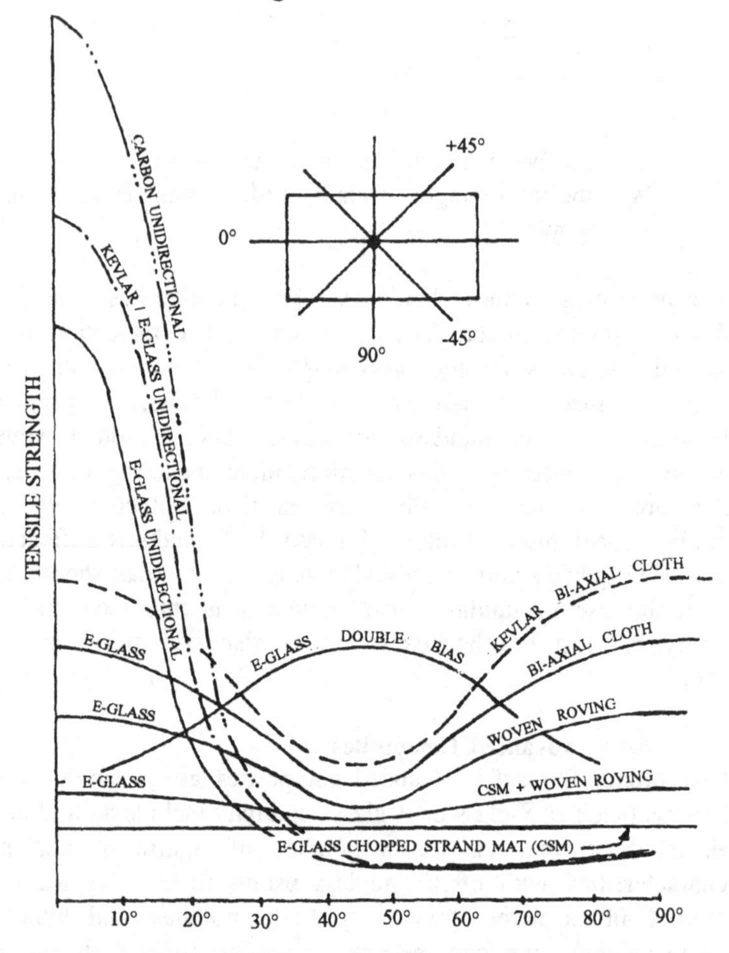

Figure 8.16. Directionality of laminates.

Because of the large variety of mechanical properties that can be obtained, the designer must use a rational engineering approach to the selection of reinforcing materials to provide the optimum design. The designer must also use a recognised engineering standard that involves established design loadings and the assessment of strength and stiffness in at least the two principal in-plane axes of the panel. Therefore, to obviate failure when using such hybrid versions, it is essential that quality workmanship and the correct selection of laminating processes together with a quality control system are used to ensure that the final properties are achieved in the resulting construction.

8.4.4 General Considerations to be Taken to Avoid Laminate Failure

Presented below is a general list of aspects to be considered for craft design to avoid failure in the moulded hull and deck:-

- To avoid the problem of progressive delamination failure of a hull moulding, it is recommended that all woven roving hulls are avoided. The inclusion of lightweight CSM (or the use of a combination mat) will greatly enhance the inter-laminar shear properties of the laminate where wet-on-wet lay-up is not practicable.
- Sudden changes in laminate section should be avoided; otherwise hard spots may be produced which may ultimately lead to the initiation of a delamination.
- All openings in the hull (e.g. sea inlet boxes, water jet penetrations and bow thruster inlets) must have well rounded corners to alleviate stress concentrations referred to above.
- Care must be exercised in drilling GRP to ensure that excessive pressure is not used which may lead to surface laminates being damaged as the drill emerges. Similarly, where bolted connections are used, the bolts should have a suitable clearance to avoid damage as the bolts are tapped through or tightened up.
- Cut edges of all holes and penetrations should be sealed with resin to avoid the ingress of moisture which will progressively lead to deterioration of the laminate.
- Extreme care must be exercised in handling and supporting the mouldings prior to stiffening. Due to the low flexural rigidity of a new moulding considerable damage can be caused by careless demoulding particularly at knuckles/chines and corners. The use of modern release agents and adequate compressed air "blow-off" points will greatly assist in demoulding without damage. All sizeable mouldings should be given adequate temporary stiffening to ensure that they can be transported/manoeuvred prior to stiffening. If this cannot be arranged then the mould should be re-designed to allow the moulding to be stiffened in the mould before removal.
- Many causes of failure in laminates can be attributed to

laminating procedures and to cleanliness. Careless mixing of resin, accidental inclusions (nails/splinters/dust etc.), spillages of materials (waxes, water, cleaning agents) can all have a dramatic effect on the final properties of the resultant laminate.

- Considerable care must be exercised in the repair of fire damaged laminates. Following a fire the laminate must be carefully inspected and a sample removed for mechanical testing. Extreme temperatures will cause the resin to reach its SADT (self accelerated decomposition temperature) and the laminate properties will be severely reduced. Surface layers need to be carefully removed and the laminate re-built to the original thickness. Additional covering plies should also be considered.

8.5 CORE MATERIALS FOR SANDWICH CONSTRUCTION

Recent years have seen an increasing variety of core materials available for sandwich construction. The range of materials available includes end-grain balsa and a number of PVC cross-link and linear structural foams which have been successfully developed for both standard and advanced designs.

Most of these materials have varying characteristics as is evident from table 8.2. Therefore to obviate failure in service, careful selection for a particular application must be made in view of their mechanical and physical properties for the intended design.

Core materials should have adequate shear strength to carry panel shear loads and adequate compressive and shear moduli to prevent the laminate skins from buckling or wrinkling. Additionally, particular attention must be given to certain foam cores which have through thickness compression properties lower than their shear strength. The listed materials give the properties which are generally adopted for calculation purposes.

In general, the shear strengths of the listed cores are fairly consistent in all through-thickness planes and their shear strengths appear to be negligibly affected by thickness for the range of materials generally used in yacht and small craft designs. These characteristics greatly simplify both the structural design and construction. Great care should be taken in considering other core materials as they may or may not have the necessary properties and characteristics, not only those referred to but the possibility that even test established properties may degrade over a period of time in

Table 8.2. Typical sandwich care materials.

TYPICAL SANDWICH CORE PROPERTIES	E. G. Balsa	Marine Ply	Firet	Airex R62.80	Divinycell H60	Divinycell H80	Divinycell H100	Plasticell D55	Plasticell D75	Plasticell D100
Density (Kg/m^3)	140	670	650	80	60	80	100	55	75	100
Shear Modulus (N/mm^2)	129	500	68	26	15	21	34	8	21	27
Shear Strength (N/mm^2)	1.64	7.59	1.11	0.92	0.60	0.93	1.15	0.49	0.95	1.16
Eff. Wb. Ten. Mod. (N/mm^2)	0	3175	350	0	0	0	0	0	0	0

service because of possible chemical instability of the core material.

When designing sandwich laminates however, it should be remembered that the shear and compressive properties published for foam core materials are frequently associated with a percentage of permanent deformation and consequently, to prevent deterioration of the laminate in-service, the factor of safety selected for the core materials must take this into consideration.

Very little use has been made of Nomex core materials with the exception of a few highly advanced designs. Nomex provides very high strength to weight ratios but considerable care must be used when using these characteristics in both design and construction.

Recent months have seen considerable interest in the adoption of new "coremat type" materials. Typical materials are Spheretex, Trevira and U-Picamat however, whilst these materials are advertised as core materials, it is recommended that since the ratio of skin thickness to core thickness is almost equal, the resulting laminate does not behave as a true sandwich construction and therefore, it is recommended that they should be designed into a structure as a weak reinforcement.

8.6 PRECAUTIONS AGAINST FAILURE IN HULL STRUCTURAL DESIGN

8.6.1 General Introduction

In sandwich construction the core thickness is usually governed by its limiting shear strength or the maximum panel size for the core thickness proposed.

The approach to hull structural design is clearly laid out in the Classification Society Rules. However, local mechanical impact loads on the outer skin should also be considered, usually in terms of providing a minimum number of plies or minimum outer skin thickness dependent upon the craft's size. In larger craft, hull girder strength should be considered even though it is unlikely to govern the hull laminate design. With advanced composites, these design considerations have become even more relevant.

Generally with standard composites strength and stiffness of the shell and deck laminates do not vary appreciably in the 0-90° axes of the panel and the strength and stiffness can be obtained with the methods used for isotropic materials such as aluminium or steel. With advanced composites the cured laminate properties of the shell and deck laminates in the 0-90° axis should be determined. Where they differ appreciably, consideration should be given to maintaining suitable panel strength and stiffness in these two axes.

In addition, the strength and stiffness of the laminate along the length of the craft should be adequate for the hull girder strength whilst in the 0-90° axes those parts of the shell or deck forming the flanges of internal stiffeners such as frames, longitudinal girders or beams should have adequate strength and stiffness in the direction of that particular structural member. Laminates should also be reinforced for local loads and local stress concentrations.

For the most efficient use of material, the use of unidirectional material should extend down the web so that full strength can be used without over stressing the adjacent laminate. The unidirectional plies on the crown should be generally protected by outer plies of CSM/woven rovings.

8.6.2 General Considerations to be Taken to Avoid Hull Structural Failure

Presented below is a general list of some of the items to be considered in the design of a hull and deck:-

- Sandwich construction is not recommended for the hulls of craft, particularly where the speed of the craft gives rise to significant impact pressures on the bottom and bow structure. Where sandwich constructions are used for the side shell of a craft it is strongly recommended that web ties are included. To be effective, these should be placed in the span direction of the panel.
- For chines/spray rails to be effective span points or stiffeners they should be both reinforced with local additional plies and should be filled with high density structural foam and overlaid with a number of plies of reinforcements. It should be remembered that where spray rails are fitted they will attract load even when positioned in way of the centre panels and therefore, it is recommended that calculations should generally be carried out to determine the design of the spray rails.
- Sudden changes of stiffener section should be avoided to reduce the possibility of local stress concentrations, i.e. "hard spots".
- To ensure continuity of longitudinal and transverse structure it is recommended that shell longitudinals are approximately one half of the depth of the web frames. The webs of frames should be checked to ensure that depth to thickness ratio will

not cause web buckling or shear problems.

The height of web frames can be significantly reduced by the inclusion of unidirectional rovings in the table or by increasing the weight of reinforcement in the table by overlapping the individual web reinforcements from each side, see figure 8.17. However, local stiffness of the hull must be taken into consideration with the global structure.

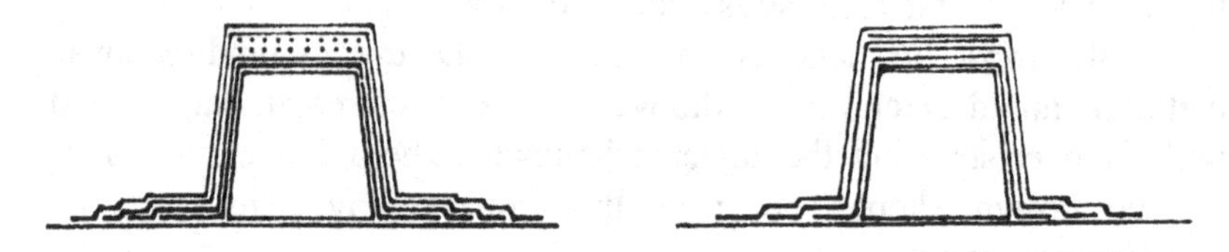

Figure 8.17. Top hat alternative configurations.

- Service experience has shown the need for additional reinforcement in the transom boundary which should progressively extend up the transom. In this respect, Classification Society Rules generally require twice the required bottom laminate in this area.
- To provide adequate protection for fresh water and diesel oil tanks the tank should be lined with the equivalent of 1800 g/m^2 CSM to give an effective water/oil barrier.
- For general guidance the thickness of plywood bulkheads can be based on a function of the length of craft ie $t_b = 1.2 L_s$ (mm) and should be bonded in with a weight of reinforcement equivalent to $112 t_b$ gram/m^2 of CSM reinforcement.

  It is essential that the boundary angle is adequately prepared and that a CSM is laid first to provide a good bond to the shell.
- Considerable care must be exercised in the correct selection of the type and grade of foam when it is used as a structural core for a deck application. Deck temperatures in hot climates may rise to a maximum temperature of 90°C depending upon colour. A foam with adequate strength retention at this temperature is essential.
- Foam cores should be carefully bonded and shear ties used where large panels are proposed. Where local loads are

imposed on sandwich construction decks the core should be either:-

a) tapered down to single skin, or
b) replaced with bonded in plywood panels with chamfered edges

- With the introduction of high moduli fibres there is a danger to concentrate on the required modulus of a stiffener without due regard to the level of shear stress in the web. A number of recent failures have occurred where web failure due to shear has been found. Figure 8,18 shows the shear distribution in the web of an idealised half section of a top hat stiffener.

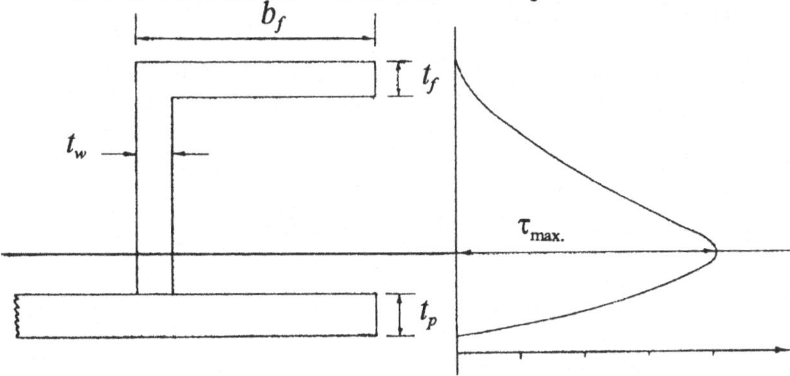

Figure 8.18. Typical shear stress distribution in idealised top hat stiffener.

8.7 DEVELOPMENTS

Recent testing of the fatigue characteristics of Kevlar, Glass and combinations of these materials has been undertaken by Lloyd's Register for Du Pont de Nemours to study the fatigue of aramid hybrid materials for advanced craft designs.

The results of this work [5] show that the ultimate load carrying capability of laminates containing aramid fibres can be considerably greater than that of glass laminates. However, to enable a significant reduction in laminate thickness to be obtained, the design of the outer layer(s), which for marine applications must not sustain surface fractures, will require to be specially considered.

8.8 REFERENCES

1] Norwood, L. S., "Blister Formation in Glass Reinforced

Plastics: Prevention Rather than Cure", Proc. 1st Intl. Conf. *Polymers in a Marine Environment*, IMarE, London 1984.

2] Tipping, G., "Water Resistant Systems for the Marine Industry", Proc. Reinf. Pl. Cong., BPF, London, 1986.

3] Tipping, G., "Osmosis: The Concept of `Total Control' Part 2", Reinf. Pl., November 1987.

4] Currey, R., "Fibre Reinforced Plastic Sailing Yachts - Some Aspects of Structural Design", Proc. 9th Chesapeake Sailing Yacht Symp., 1989.

5] Howson, J.C., Rymill, R.J., Pinzelli, R.F., "Fatigue Performance of Marine Laminates Reinforced with Kevlar Aramid Fibre", Proc. Intl. Conf. *Journees Europeennes des Composites*, Paris, April, 1992.

9 RESPONSE OF SANDWICH STRUCTURES TO SLAMMING AND IMPACT LOADS

9.1 INTRODUCTION

While FRP sandwich has been used in racing and pleasure craft for some considerable time, it is only since the early and mid 1980s that this material has established itself as a major construction material for hulls, decks and superstructures of larger vessels such as high speed passenger ferries. Norwegian and Swedish ship builders have been in the forefront of this development, and it is significant that Norway is currently building nine new mine countermeasures vessels in GRP sandwich (including the shock-exposed areas of the hulls).

In spite of the high degree of confidence attained in the use of FRP sandwich, there have been a number of failures, due to various causes, and it is important to be clear about the problems that can arise if the structure is not correctly designed and built.

In a sandwich panel the skin laminates and the core are of different materials and these have to be fixed together by some form of adhesive bonding. This presents a series of challenges, with respect to both workmanship and materials selection. It has also become extremely important to understand properly the loadings and mechanisms involved, particularly in relation to slamming and to impact with solid objects.

In the autumn of 1988 extremely severe weather was experienced around parts of the Norwegian coast. Several vessels experienced damage - not only ones built in GRP, but there was a relatively high incidence of GRP sandwich failures.

The picture seen typically was a large area of delamination - or, more correctly, separation of the outer skin from the core, with extensive destruction of the PVC foam core, figure 9.1. This damage had apparently been caused by severe slamming loads applied to the outer surface. At first attention was focused almost entirely on the properties of the core material. Samples taken from one of the affected vessels showed that the core had a lower fracture elongation than normally expected. However, other vessels with more ductile cores had also suffered damage, so this was clearly not the whole story.

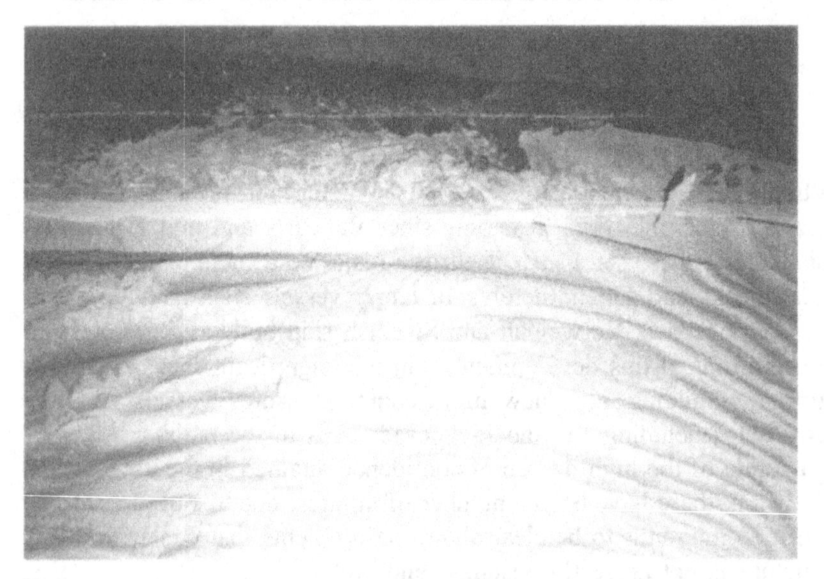

Figure 9.1. Slamming damage to sandwich hull, after removal of outer skin.

In this Chapter the problem of slamming on GRP sandwich panels is examined in more detail, and it will be shown that there were a number of factors that probably contributed to these failures. It will also be shown that, with correct attention to the definition of slamming loads, structural response, material selection and core fabrication, it is possible to eliminate, or at least greatly reduce, the incidence of such failures.

There has also been a series of incidents where impact with solid objects in the water has apparently led to progressive failure of the sandwich hull. This is a different problem. Finally, ways in which this type of failure may be combated will also be explored.

9.2 SLAMMING LOADS

In designing against slamming it is usual to analyse a panel under a uniformly distributed static pressure that is assumed to represent the effect of a characteristic slamming event. In reality the slamming pressure is not uniform: it is a pressure pulse that typically starts at one edge of a panel and works its way across to the other edge, as shown in figure 9.2. These are pressure signals from six transducers placed in a line across a sandwich panel forming one bottom panel of a V-shaped hull model [1]. Pressure transducer P1 was close to the keel and P6 close to the chine. The model was dropped into calm water from a height of 7 m. The orientation was such that the angle β between the panel and the water surface was 30°. Figure 9.3 shows the corresponding signals from a test with drop height 2.2 m when the impact angle β was reduced to 5°.

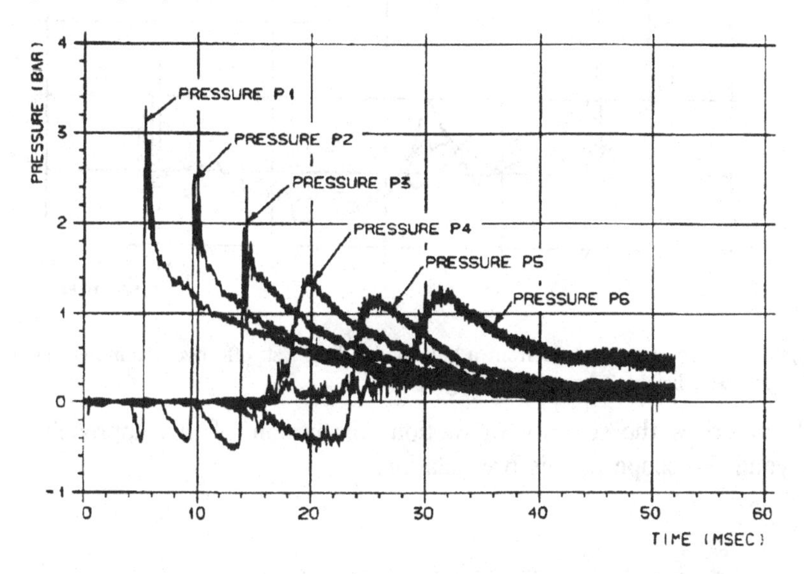

Figure 9.2. Pressure signals from a hull model drop test: effective slamming angle β = 30°, drop height 7 m.

The peak pressure in such a slamming pulse is approximately proportional to V^2, where V is the velocity of the panel normal to the water surface. It is also a function of the effective slamming angle β. Theoretical studies of the water entry problem suggest that the peak pressure becomes infinite as β tends to zero. However, in practice as β is reduced below about 3-4° air becomes trapped between the panel and water surfaces and cushions the impact. The relationship between maximum pressure and β is thus of the type shown in figure 9.4.

In some cases the peak pressure has little relevance because it is of extremely short duration and, at any given instant, covers an extremely small part of the panel, as illustrated in figure 9.3. Thus the *effective* slamming pressure is also a function of the panel size. Classification rules reflect the dependence on β and on the design area considered, A. The penetration velocity V is typically accounted for by referring to a design vertical acceleration for the vessel, which

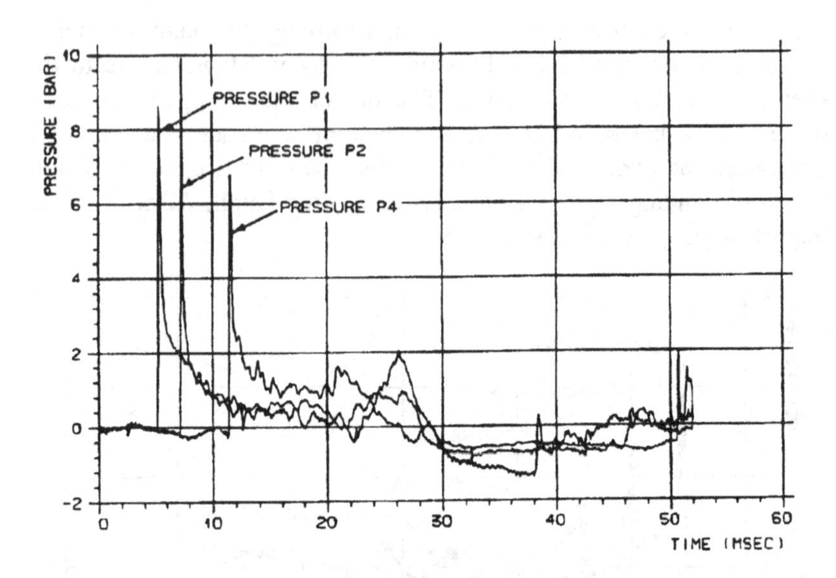

Figure 9.3. Pressure signals from a hull model drop test: effective slamming angle $\beta = 5°$, drop height 2.2 m (filtered signals).

characterises the severity of motion; discussion of this approach is beyond the scope of this presentation.

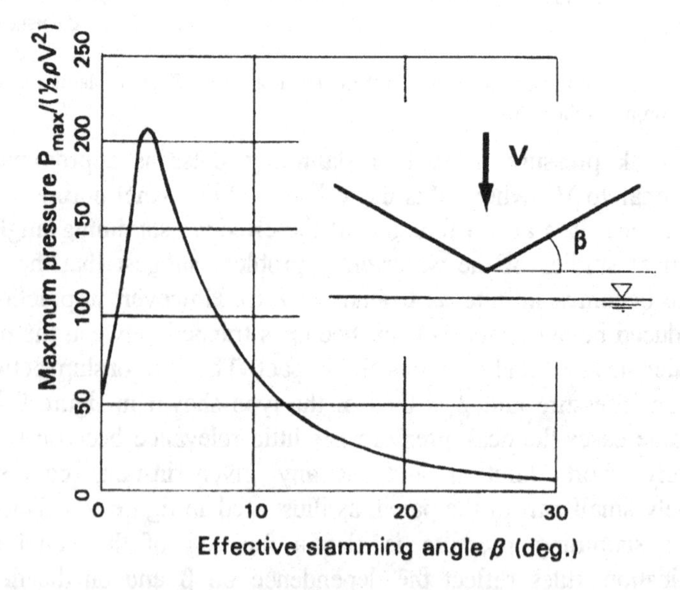

Figure 9.4. Relationship between peak pressure and effective slamming angle for penetration of a water surface by a wedge.

A number of research studies have been carried out into the slamming loads experienced by bodies of various types as they penetrate a water surface. However, not enough attention has been paid to the measurement of induced stresses and deformations in the structure, which are the real measure of the effective slamming loads. In 1989, as part of a Norwegian collaborative research and development programme into fast vessels, DNV started investigations in this field with the specific aims of finding out:

- how GRP sandwich panels respond to individual slamming events
- whether there are differences in response as compared to stiffened aluminium hull structures
- whether the panel flexibility has an influence on the slamming pressures
- how GRP sandwich panels (and particularly PVC foam core materials) respond to repeated slamming loads.

9.3 RESPONSE OF SANDWICH PANELS TO SLAMMING LOADS

The slamming pressure histories in figures 9.2 and 9.3 show that slamming loads often resemble travelling concentrated loads rather than uniform static pressures. Figures 9.5 and 9.6 show shear strain histories actually measured at successive positions inside the core of the panel on which these pressure histories were measured. Strain gauge position SB is close to the keel while SI is close to the chine. The shear strains are also consistent with the passing of a travelling pressure pulse.

Figure 9.7 shows damage to one of the bottom panels after conclusion of the tests. There was a clearly visible area of delamination near the upper edge of the panel. When the panel was subsequently cut up for inspection the core was found to have suffered extensive shear failure in this region. There was also some damage near the lower edge, adjacent to the keel. Figure 9.8 is a close-up view of an area of extensive core shear failure and delamination on a second sandwich model. The important thing to recognise here is that core shear failure has come first, and the delamination is a secondary effect.

The bottom panels were designed in accordance with the 1985 DNV Rules for Classification of High Speed Light Craft, for a given equivalent static pressure. It was found during the testing that the

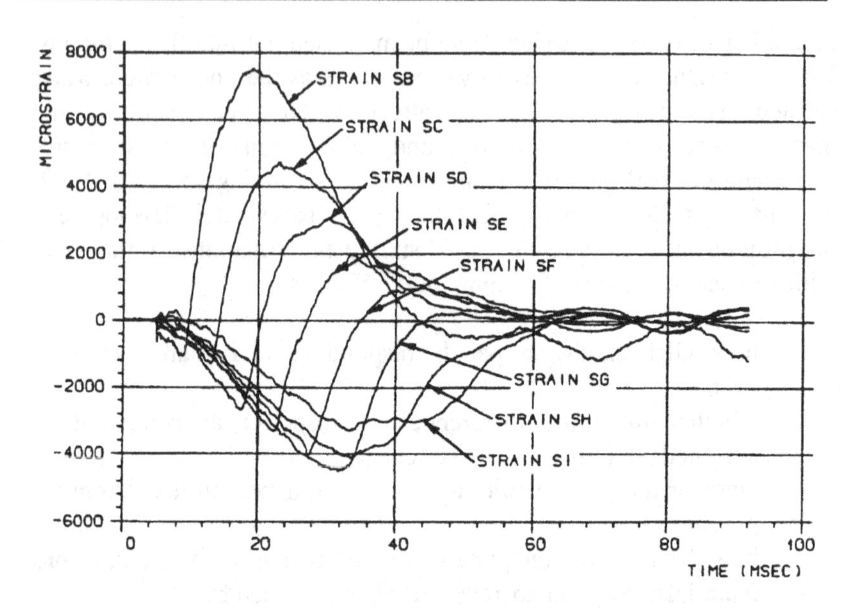

Figure 9.5. Shear strain histories measured at successive positions in sandwich core of hull drop test model: effective slamming angle $\beta = 30°$, drop height 7 m.

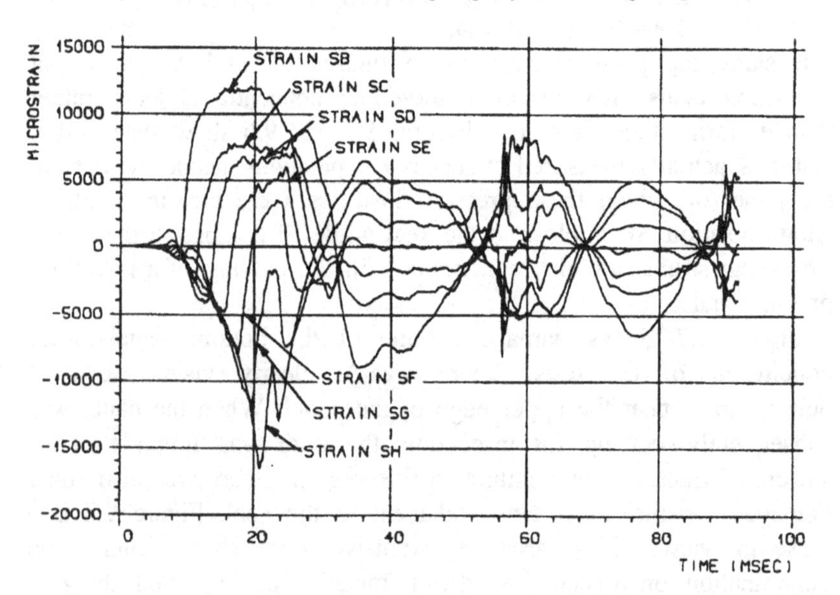

Figure 9.6. Shear strain histories measured at successive positions in sandwich core of hull drop test model: effective slamming angle $\beta = 5°$, drop height 2.2 m.

laminates could withstand considerably higher slamming loads than the core. One reason is that the allowable shear stress for sandwich cores was an appreciably higher fraction of the ultimate shear stress than the allowable tensile stress in the skin laminates was of the ultimate tensile strength. However, part of the difference was due to the special character of the slamming loads.

Figure 9.7. Damage to sandwich bottom panel in drop test model.

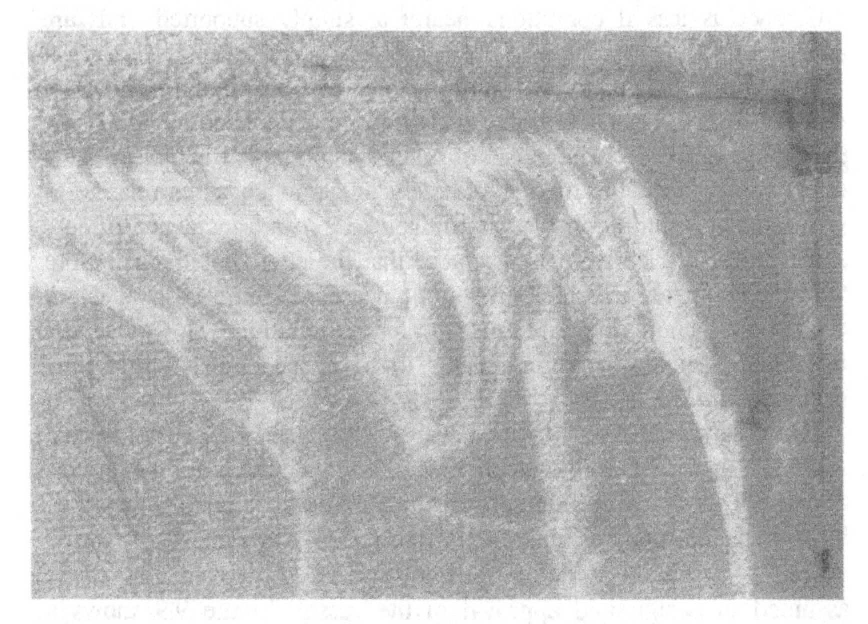

Figure 9.8. Damage to sandwich bottom panel in drop test model.

Formulae for effective slamming pressure have been established in the past on the basis of steel and aluminium plating. These are essentially subjected to bending stresses and deformation. Consider a long bottom panel, of width b, which can be represented by a beam of length b placed in the transverse direction. It is assumed that the panel has restrained edges so that the beam must be considered with built-in ends. Suppose the slamming load can be represented by a concentrated, travelling load W. The maximum bending moment induced is Wb/8, when the load is at the centre. If a uniformly distributed load of intensity w is applied, this induces a maximum bending moment $wb^2/12$. Thus the two loadings induce equal maximum bending moments if w = 1.5 W/b, so that a uniformly distributed load 1.5 W/b may be considered equivalent to the concentrated load W.

Now consider the shear forces set up by the two "equivalent" loadings. The worst position for the concentrated load W is now at one end of the beam. Here it induces a maximum shear force of W. The uniformly distributed load induces a maximum shear force wb/2 = 0.75 W. Thus the "equivalent" uniformly distributed load gives a misleading impression of the maximum shear force induced the concentrated load, underestimating it by 25%. (However, the difference is less if conditions nearer to simply supported ends are assumed.)

In a similar way equivalent uniform pressures for slamming that are derived from experience with panels that undergo bending are lower than those derived on the basis of maintaining the correct shear. This effect has reduced the margin of safety in sandwich cores relative to skin laminates, and probably contributed to the high incidence of core shear failures. Clearly, it is desirable that the core should fail before the skin laminates in a hull structure, since core failure in itself is not catastrophic, nor does it cause water penetration. However, the difference in safety margins seems to have been much larger than originally intended.

Particularly extensive failures were experienced by coastal rescue vessels; figure 9.1 shows such a failure. In addition to the effects described above, a contributory feature to these failures appears to have been an underestimation of the loads experienced by these vessels - or their operation outside the speed/wave height range assumed in design and approval of the vessel. Figure 9.9 shows a vertical acceleration history measured close to the LCG of a coastal rescue vessel experiencing a severe slam while travelling at about 18 knots in waves of about 4 m [1]. This shows a peak vertical acceleration of just over 6 g, where g is the gravitational acceleration. This has become more of a problem with patrol and rescue vessels in which the crew are provided with seating having advanced air-suspension systems: the loading experienced by the vessel is no longer limited directly by what the crew can tolerate.

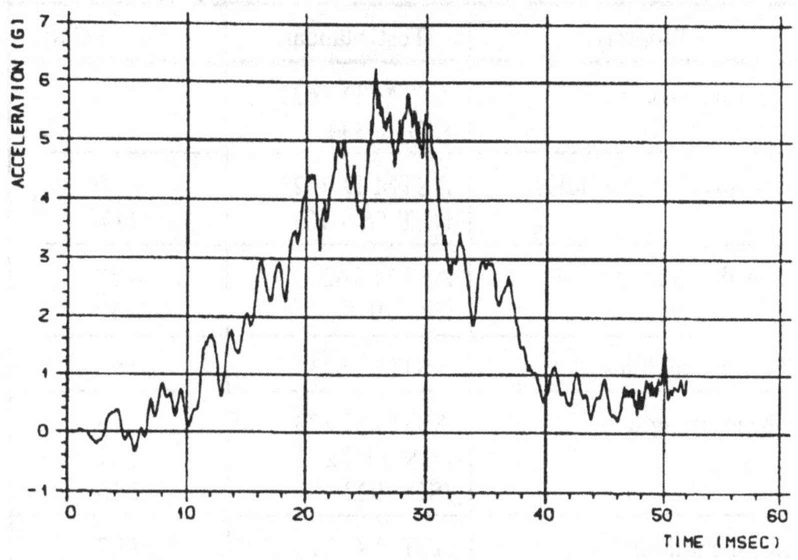

Figure 9.9. Vertical acceleration measured close to the LCG of a coastal rescue vessel.

9.4 MATERIAL PROPERTIES AND TEST METHODS; MANUFACTURERS' DATA SHEETS

A feature that has come to light in the discussions about sandwich failures is a variation of practice in the specification of core material properties in manufacturers' data sheets. A major reason for this is that, for rigid and semi-rigid foam materials, the properties can vary greatly according to test method. This problem has been highlighted by Polimex in a data sheet for Klegecell cross-linked PVC core, from which the data in table 9.1 are extracted. While the two commonly used block shear tests have here given similar values for shear strength, some care has to be taken in interpreting shear modulus data from these tests (though the use of thickened steel fittings usually reduces the variation considerably below that shown here). Similarly, compressive strength does not vary too much according to test method but compressive modulus figures can vary by a factor of two or more according to the test method used.

Table 9.1. Mechanical properties for a cross-linked PVC foam core material by different test methods (from Polimex data sheet).

Property	Test Standard	Value (MPa)
Compressive strength	ASTM D 1621 ISO R 844	3.06 3.27
Compressive modulus	ASTM D 1621 NFT 56 - 101	197 117
Tensile strength	ASTM 1623 ISO 1926	4.38 3.57
Tensile modulus	ASTM 1623	180
Shear strength	ASTM C 273 DIN 53422 ISO 1922	2.18 2.41 2.18
Shear modulus	ASTM C 273 DIN 53422 ISO 1922	55.7 41.2 31.2

In compression testing, to avoid buckling, the test specimen is usually rather short and fat. The cross-section is either square or circular. The specimen is placed between two steel plattens which partially restrain deformations in the direction transverse to the loading axis.

If we consider the extreme case of a compressed thin disc with large diameter such that displacements in the radial direction are totally prevented, the apparent elastic modulus $E' = \sigma/\varepsilon$ is given by

$$E' = E(1-\nu) / [(1+\nu)(1-2\nu)] \tag{9.1}$$

PVC foam core materials have Poisson's ratio values nearer to 0.4 than 0.3. If $\nu = 0.4$, $E' = 2.14$ E.

In reality a compression test will lead to a Young's modulus estimate that lies somewhere between E and E′, but where in that range will depend on the specimen geometry and the contact conditions at the specimen ends. Fortunately the compressive strength is not as sensitive to test method as the modulus.

Further factors that can lead to confusion in core material properties are (a) variations in density (and hence in mechanical properties) through the thickness of the sheet of core material and (b) anisotropy giving different properties in the thickness direction from those in the plane of the sheet.

Yet another feature is that it has been normal practice for manufacturers to specify typical or average values for density and mechanical properties, rather than minimum values or mean value minus a given number of standard deviations, as is common for metallic materials. In the context of DNV Rules, this aspect is being eliminated as manufacturers are now required to specify minimum values of critical mechanical properties when obtaining type approval, and these minimum values are stated on the approval certificate. Nevertheless, care is still needed in the interpretation of some mechanical properties of foam core materials.

9.5 EFFECT OF SINGLE AND REPEATED SLAMMING LOADS ON FOAM CORE MATERIALS

Several questions present themselves when foam core materials are considered under slamming loading:

- How is the static stress-strain curve modified by strain rate effects when subjected to typical slamming pulses?
- What is the resistance of such materials to repeated slamming loads; in particular, is the fatigue life for repeated slamming different from that for sinusoidal loading with the same amplitude?
- If the core experiences a single overload into the nonlinear range, how does this affect the subsequent properties (strength, fracture strain, fatigue life)?
- How do adhesive joints in the core affect the response to single and repeated slamming loads?

The answers to these questions are emerging from research that has been going on during the past few years at DNV and other establishments. At DNV extensive testing has been performed on sandwich beams with cross-linked and linear PVC cores under four-point bending, figure 9.10. The region between the load points and the supports is in a state close to pure shear, and the skin laminates are made thick enough to eliminate laminate failure. The loading has been:

- quasi-static increasing/decreasing
- simulated slamming, with loading histories designed to produce shear strain histories similar to those observed in slamming drop tests and full-scale measurements on a sandwich vessel in service (max./min. stress ratio, R = 0.1)
- repeated slamming loads
- conventional sinusoidal repeated loading with stress ratio R = 0.1

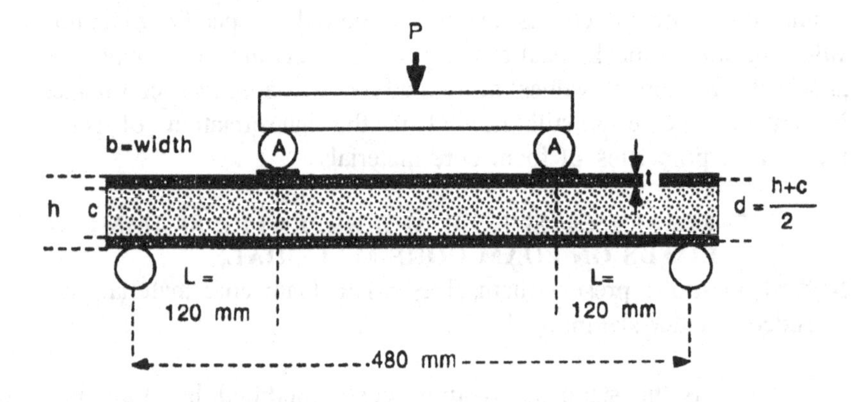

Figure 9.10. Test set-up for four-point bending of sandwich beams.

Figure 9.11 shows a set of curves of shear stress plotted against beam deflection due to slamming loading, compared with static loading, for a cross-linked PVC foam core of medium density. The shear modulus does not appear to be significantly different from the static value, but the proportionality limit and the ultimate strength are both enhanced while the fracture strain is reduced. Thus the material can withstand higher shear stresses in a slamming situation than the static properties indicate, but the behaviour is more brittle indicating that the influence of defects may be greater and the energy-absorbing properties may be less favourable. However, it must be noted this reduction of ductility does not occur for all PVC foams.

As far as repeated slamming loads are concerned, no significant difference has been found between fatigue life for rapid slamming loads and sinusoidal loading carried out at normal loading rates.

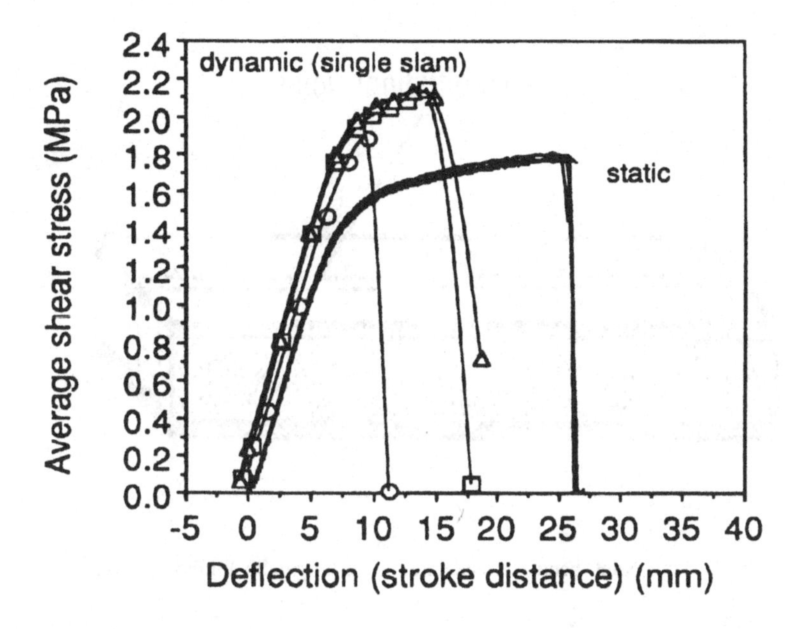

Figure 9.11. Shear stress - deflection curves for core of sandwich beam subjected to simulated slamming loading, compared with static loading.

However, single overloads into the nonlinear range may have an effect on subsequent fatigue life.

A major result of the research at DNV has been the observation that longitudinal adhesive joints in cross-linked PVC cores, figure 9.12, can influence both short and long-term slamming response. Some core adhesives are considerably stiffer than the core materials they are expected to join. Thus, when loaded, they develop appreciably higher shear stresses than the surrounding foam core. With low ductility, the adhesive cracks at an early stage in the loading, figure 9.13. If the beam is subjected to one or more slamming loads, so that the foam core takes on the more brittle, less damage tolerant behaviour described earlier, the crack in the adhesive can initiate cracking and cause premature failure in the surrounding core. Thus it is important to match the properties of the adhesive and the core as closely as possible and to ensure that the combination is sufficiently damage tolerant. However, tests with relatively brittle adhesives in a linear PVC foam core have suggested that this core material has sufficient damage tolerance itself not to be affected by crack initiation in the adhesive. A high-density cross-linked PVC

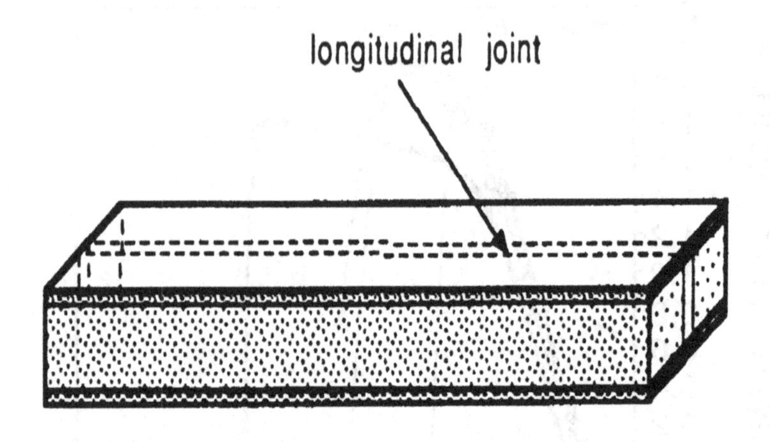

Figure 9.12. Four-point bend test specimen with longitudinal joint in core.

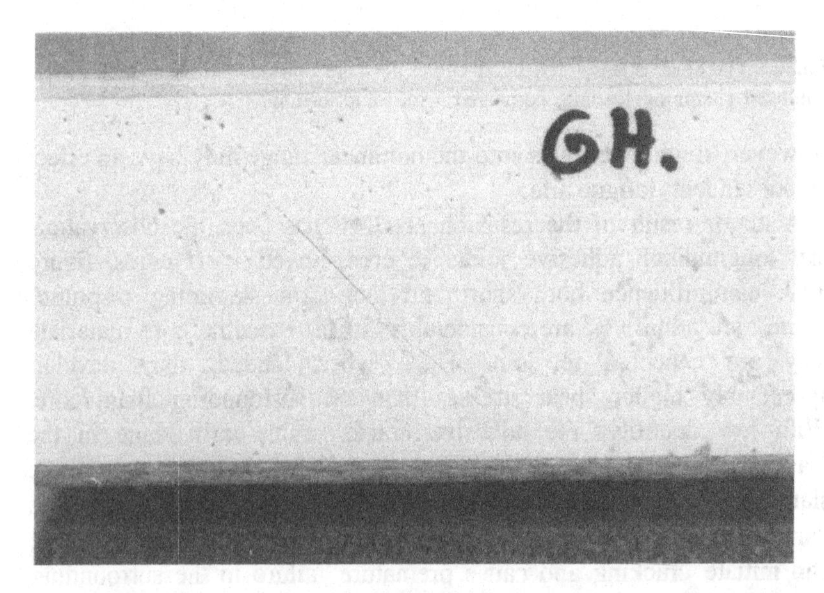

Figure 9.13. Crack formed in longitudinal adhesive joint in core of four-point bend test specimen.

foam was also found to be less sensitive to adhesive cracking than a low-density one.

A further complication is that both core adhesives and core

materials themselves may be temperature-sensitive, with much more brittle behaviour at 0°C than at 20°C.

The problem of crack initiation in core joints is now dealt with in the DNV Rules by the inclusion of requirements for fracture elongation in core materials and adhesives and for testing of beams containing longitudinal joints.

9.6 IMPACT WITH SOLID OBJECTS

Resistance of high speed vessels to impact is a topic of major concern. This applies to everything from major collision with a rock face or another vessel to striking a small submerged object in the water. Recently there have been incidents in which what is believed to have been a rather minor impact with a submerged object appears to have led to a major separation of the outer skin from the sandwich core over a large part of the sandwich hull, see figures 9.14, 9.15.

Figure 9.14. Extensive skin/core separation on a catamaran hull initiated by local impact.

When a solid object hits a sandwich panel with a thin outer skin laminate, the core suffers local crushing and the skin separates from the core over all or part of the area directly involved in the impact. If the outer skin is penetrated, even very slightly, a hydraulic pressure can build up in the small gap between skin and core. Over a period of time this pressure may cause the zone of separation to spread over a larger area of the hull - in some cases over the complete hull.

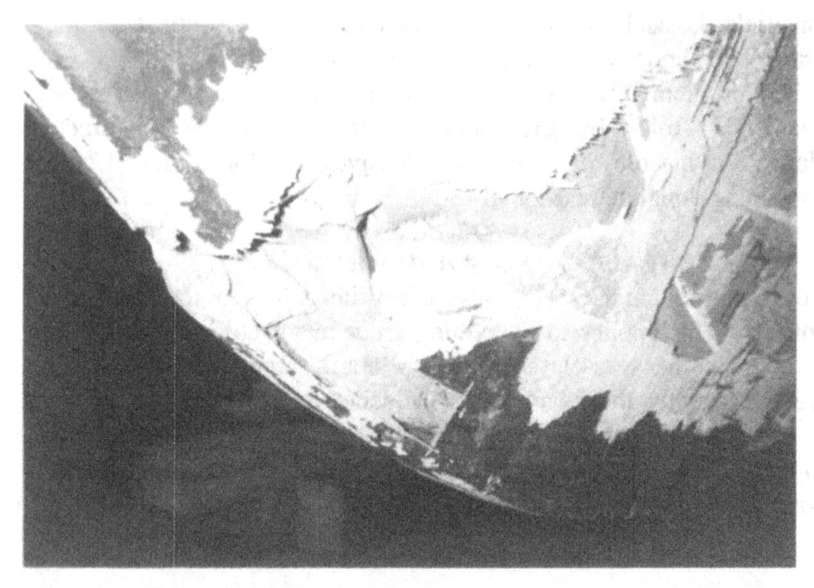

Figure 9.15. Point of initial impact damage.

Figure 12.11 in Chapter 12 shows two means of combating this type of failure, taken from the DNV Tentative Rules for High Speed and Light Craft [5]. Thickening of the outer skin laminate in the most exposed part of the hull will limit the local damage and reduce the likelihood of penetration of the outer skin. Use of a stronger, denser core material in this region may also help in this respect. Tying together the inner and outer skins as shown will ensure that laminate separation, if it does occur, is effectively arrested.

Associated with the problem of limiting damage is that of detecting it at an early stage. This is also a major topic of research.

9.7 REFERENCES

1] Hayman, B., Haug, T., Valsgård, S. "Response of Fast Craft Hull Structures to Slamming Loads". Proc. 1st Intl. Conf. *Fast Sea Transportation,* Trondheim, 1991.

2] Hayman, B., Haug, T., Valsgård, S. "Slamming Drop Tests on a GRP Sandwich Hull Model". Proc. 2nd Intl. Conf. *Sandwich Construction,* Gainesville, Florida, 1992.

3] Buene, L., Echtermeyer, A.T., Sund, O.E., Nygård, M., Hayman, B., "Assessment of Long Term Effects of Slamming Loads on FRP Sandwich Panels". Proc. 1st Intl. Conf. *Fast Sea Transportation,* Trondheim, 1991.

4] Buene, L., Echtermeyer, A.T., Hayman, B., Sund, O.E., Engh, B., "Shear Properties of GRP Sandwich Beams Subject to Slamming Loads". Proc. 2nd Intl. Conf. *Sandwich Construction*, Gainesville, Florida, March 1992.

5] "Tentative Rules for Classification of High Speed and Light Craft", Det Norske Veritas, 1991.

10 FATIGUE CHARACTERISTICS

10.1 INTRODUCTION

Fatigue may be described as a process which causes damage in material and structure under fluctuating loads of a magnitude much less than the static failure load. The accumulated damage may result in a gradual and significant decrease of mechanical properties such as strength and stiffness, in crack growth and finally into complete failure or collapse.

The speed of the fatigue process will be governed primarily by the magnitude of the fluctuations of the load or deformation cycles, commonly referred to as stress or strain range. Another load effect parameter is the height of the mean or peak level of the load cycle, although this parameter is of less importance as compared to the stress range.

The total number of load cycles which can be endured by the material or structure before fatigue failure is called fatigue life or endurance. Fatigue failure can be defined either as total collapse or as loss of adequate strength or stiffness. The maximum cyclic load or stress range a material or structure can withstand for a given fatigue life is called fatigue strength, while the maximum cyclic load that can be resisted indefinitely without failure is known as (lower) fatigue limit.

The relationship between the fatigue strength and the fatigue life is mostly presented in diagrams by means of S-N or Wöhler curves for constant amplitude loading, see figure 10.1. The horizontal axis gives the number of cycles N on a log-scale, while in vertical direction the fatigue strength might be given on a log-scale or on a linear basis. The fatigue strength will be presented preferably in the form of stress range $\Delta\sigma$ or double amplitude of the stress variation. However, sometimes the fatigue strength is represented by the load amplitude or even by the maximum of the load cycle. The most interesting part of the curves between 10^3 and 5×10^6 cycles can often be represented by a straight line of the expression:

$$\log (\Delta\sigma) = c_1 - c_2 \log N \qquad (10.1)$$

or: $\Delta\sigma = c_1 - c_2 \log N$,

where c_1 and c_2 are constants.

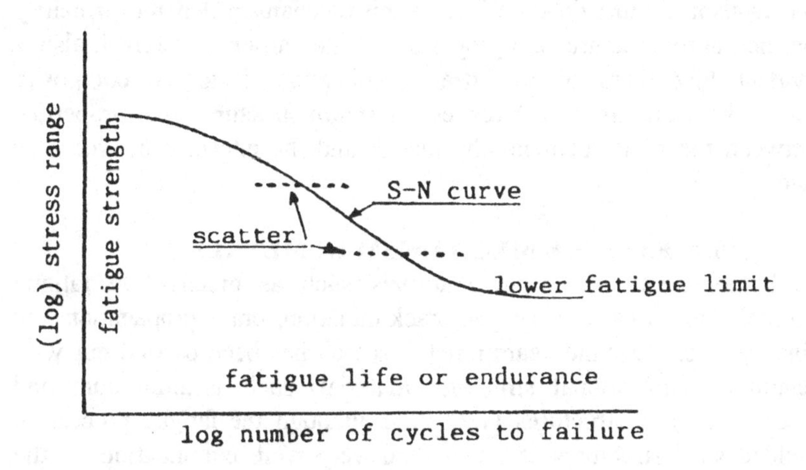

Figure 10.1. Fatigue strength versus fatigue life: S-N curve.

Normally, the first formula will show the best fit with test results at high cycle fatigue, while at low cycle, high stress fatigue levels the second formula may be preferred. A series of such curves may need to be produced for the different types of loading which can occur, that is with different levels of mean stress.

The mean S-N curve represents a 50/50 chance of failure and a component designed on the basis of a mean curve has an equal chance of failure or survival. In design, a greater chance of survival is usually required and the design S-N curve falls below the mean.

Because all marine structures are subjected to varying loads under normal working conditions, fatigue is a common cause of damage and failure in service. Designing against fatigue failure and prediction of fatigue life is difficult and complex due to the many and often not well-known influencing parameters. The fatigue behaviour will be influenced by micro-structure and material properties, dimensions and geometry, production aspects, loading conditions and load history, environment and lastly by the mutual influence of all these parameters. To determine the influence of these parameters on the fatigue behaviour of materials and structural components laboratory tests on small standard specimens and large scale structural components are unavoidable. By doing this, it is self-evident that the varying parameters are brought into as good agreement with service conditions as it is possible to obtain.

However, before doing fatigue testing, analysing the results and predicting fatigue behaviour, it is desirable to look at the fatigue mechanism in more detail. Whereas this mechanism depends primarily on the micro-structure and properties of the chosen material, it also is evident that there is an extra complicating factor in composite materials with large differences in micro-structure and properties between the reinforcement, the matrix and the interface between the two.

10.2 FATIGUE MECHANISM IN METALS

In homogeneous, isotropic materials such as metals, the fatigue process comprises three stages; crack initiation, crack propagation and final failure. Over the years much research has been carried out with regard to conventional structural materials such as aluminium and steel for marine structures [1,2]. And although the fatigue process in welded steel structures is understood very well, continuation of the research on a rather large scale is still thought to be necessary.

Crack initiation is linked to microscopic material behaviour. Cracks are initiated by localised plastic deformation as a result of dislocation motion under cyclic strain. Since the dislocation mobility is greater at a free surface than in the bulk of the material, in general, fatigue initiation is a surface phenomenon. Normally cracks will initiate at the locations with the largest stress or strain range, that is at the points of stress concentration due to discontinuities in geometry, surface roughness or defects. The term "surface" refers to outer surfaces and also to inner surfaces at embedded flaws or defects.

Under successive load cycles dislocation motions occur in different adjacent slip planes in the crystalline material, resulting in extrusions and intrusions on the surface. The formation of these intrusions is the initiation stage. Microscopic slip may occur in single grains at stresses below the general yield stress of the material. Under continued load cycling, one or more intrusions may develop into what is commonly denoted stage I crack growth. In this stage the crack follows in a shear mode the same crystallographic plane upon which slip originally took place, i.e. in a plane oriented at 45° to the maximum principal stress direction. Commonly these two stages are referred to as "initiation" or "crack nucleation".

With increasing crack depth the direction of crack growth will change rather quickly from 45° to a direction normal to the direction of the largest tensile stress. From this point the maximum principal stress, and not the shear stress, will be the driving force. The crack growth rate depends on the stress concentration at the crack tip, characterised by the stress intensity factor K, which is a function of the applied load (or nominal stress), crack length and shape, and external geometry. K may be conveniently expressed as:

$$K = \sigma \cdot \sqrt{(\pi a)} \cdot f(a/b) \qquad (10.2)$$

where a is the crack depth, b is specimen width and f(a/b) is a function of geometry of crack and structural component in relation to the external loading mode. From this expression it is clear that in general K increases with crack growth and so will the crack growth rate. The general form of the relationship between crack growth increment in a load cycle and the range of the stress intensity factor ΔK in the same load cycle is shown in figure 10.2 for crack growth in a tensile mode. For many practical cases the relation can be expressed by the so-called Paris-Erdogan crack growth law:

$$da/dn = C(\Delta K)^m \qquad (10.3)$$

where C and m are material constants which have to be determined experimentally. This relationship is the basis for the application of fracture mechanics to fatigue of steel structures.

At the upper part of the curve, at higher values of ΔK, crack growth exhibits a rapid increase of growth rate towards "infinity" by ductile tearing and/or brittle fracture at a maximum or critical value of the stress intensity factor K_{max} or K_c. At low values of ΔK, the crack growth rate goes asymptotically to zero as ΔK approaches a threshold value ΔK_{th}. This means that there is a lower fatigue limit at ΔK_{th}. The threshold effect is believed to be caused by a complicated synergism of processes and the threshold level depends on stress parameters like mean stress, residual stresses and stress interaction as well as on material properties and environmental effects.

Considering the total fatigue life N_f, it should be borne in mind that this includes an initiation stage N_i and a propagation stage N_p, the latter primarily depending on the nominal loading conditions, the former, above all, depending on local stress raising factors. At low stress levels and with smooth and undamaged surfaces, the initiation stage may occupy up to about 90% of the fatigue life. However, the

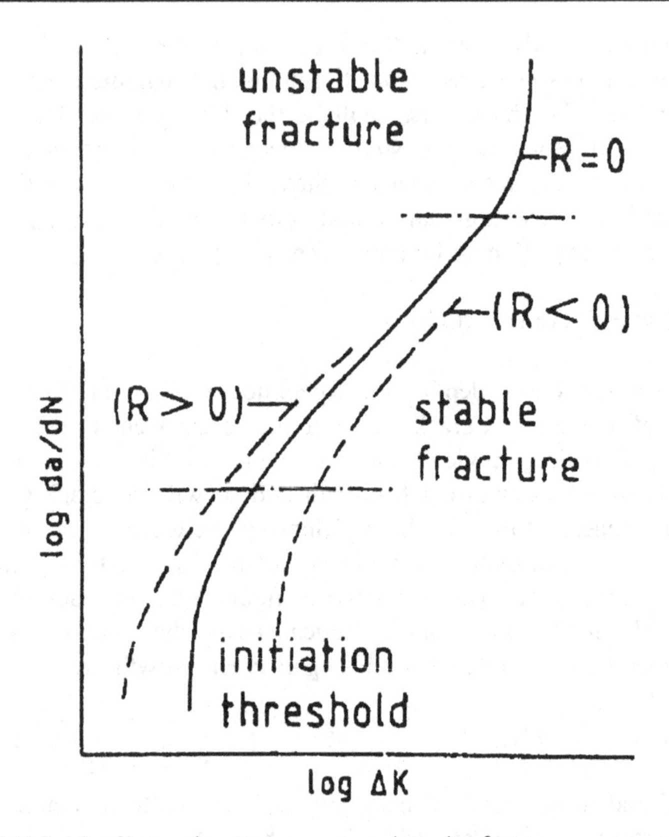

Figure 10.2. Crack growth rate versus stress intensity factor.

existence of stress concentrations due to defects, surface roughness or discontinuities, may reduce the initiation phase to an almost negligible proportion, and also the fatigue life is correspondingly shortened.

Considering the environment, this may influence the initiation stage as well as the stage of crack growth.

10.3 FATIGUE MECHANISM IN COMPOSITE MATERIALS

The mechanism of the fatigue process in composite materials or fibre reinforced plastics is quite different from the mechanism in steel and depends on the properties of matrix and reinforcement as well as on the interaction between the two [3-6]. The problem with reinforced plastics is that they are inhomogeneous, anisotropic, and rarely behave in a linear elastic fashion. Crack propagation is not as simple a process as it is in metals. Crack paths are highly complex, and a crack itself is not the only manifestation of structural damage.

10.3.1 Fibres

A common mode of failure of inorganic glass fibres is that which is termed "static fatigue" and more properly described by the metallurgical term "stress-rupture". This is, in fact, brittle creep. Failure in fibres other than glass may occur through the usual stages of initiation, propagation and final failure; the details depending upon the particular nature of the fibrous material.

10.3.2 Polymers

The polymeric matrix differs from metals in that it does not have an equivalent to stage I initiation and crack growth. Under low stress conditions cracks may initiate at the specimen surface, or internally from existing flaws, by a form of slip in crystalline polymers, or from voids formed during viscous flow. This makes crack initiation correspondingly more important.

Polymers suffer also from two other types of failure not found in metals. These are cyclic creep and thermal failure, which arise respectively under conditions of high stress and high loading frequency. Cyclic creep occurs when the load is sufficiently high to cause static creep. Thermal failure occurs when energy loss due to inherent high damping in the polymer cannot be dissipated because of the material's low thermal diffusivity, and the resulting temperature rise causes thermal softening and loss in properties. For this reason fatigue testing of polymers and FRP has to be limited to rather low frequencies, less than about 5 Hz.

Although the detailed atomic or molecular processes are quite different, the general progress of crack growth in the various types of polymeric material is seen to be very similar to that in metals.

10.3.3 Interface

Finally, the interface between matrix and reinforcement may have a significant effect on crack initiation as well as on crack propagation, depending on the bonding strength of the interface and of the orientation of fibres and principal stresses. On the one hand, stress concentrations at, or debonding of, the interface can lead to rapid crack initiation and drastically reduce the life of a matrix whose resistance to fatigue is dependant upon the difficulty of crack initiation. On the other hand the presence of interfaces may considerably slow down crack propagation by effectively blocking and changing the direction of propagation of the crack. In addition, friction developed between fibre and matrix can absorb energy needed for crack propagation.

10.3.4 Laminate

From the foregoing it is clear that there are three basic failure mechanisms in composite materials as a result of fatigue: matrix cracking, interfacial debonding and fibre breakage. These basic failure mechanisms and the interactions are described extensively by Talreja [6], and figure 10.3. A fourth basic failure mechanism as a result of matrix cracking and interfacial debonding is called delamination, which is the debonding between adjacent layers of reinforcement in multi-layer laminates and connections.

Investigations into the development of damage in chopped strand mat/polyester resin laminates under tensile and fatigue loading have been carried out by Owen [7,8] and others. The investigations showed the following sequence of degradation:

- adhesion failures between fibres and resin
- cracking of the resin
- fractures of single fibres
- total failure by a propagating macroscopic crack surrounded by numerous subcracks

There are two distinct stages in the development of damage in this material. At an early stage debonding occurs between glass and resin among transverse fibres lying perpendicular to the line of load. Under static tensile loading, debonding starts at stress values between 10 and 30% of the ultimate tensile stress. The number of debonds increases rapidly at most sites at between 30 and 70% of UTS and tends to become saturated, while at about 70% UTS level the debonding coincides with the onset of resin cracking, again perpendicular to the line of loading.

If the stress level is high enough to produce transverse fibre debonding in a fatigue test, the number of debonds builds up rapidly and tends to a constant value until the latter stages of the test when it again increases. At first the specimens become slightly opaque and show a whitening effect under tensile load, the opacity disappearing in the compressive part of the cycle. Gradually the opacity becomes permanent, showing white spots, and resin cracks can be seen. Resin cracks produced under fatigue loading are much more numerous than under static tensile loading and again accumulate more rapidly in the early stages of the testing than in the latter stages.

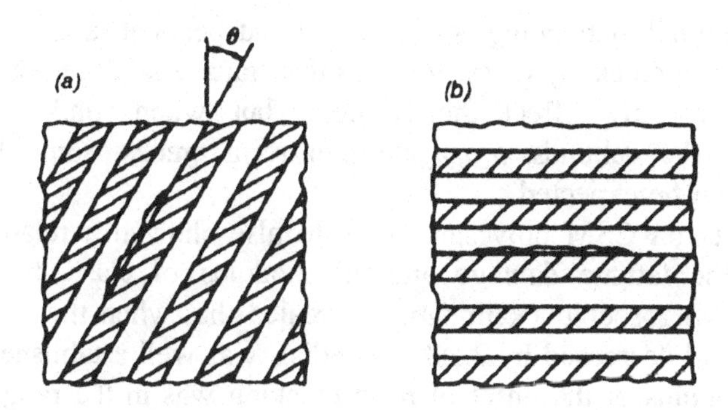

Matrix and interfacial cracking under off-axis fatigue of unidirectional composites: (a) mixed (opening and sliding) mode crack growth, $0 < \theta < 90°$ (b) opening mode crack growth (transverse fibre debonding), $\theta = 90°$.

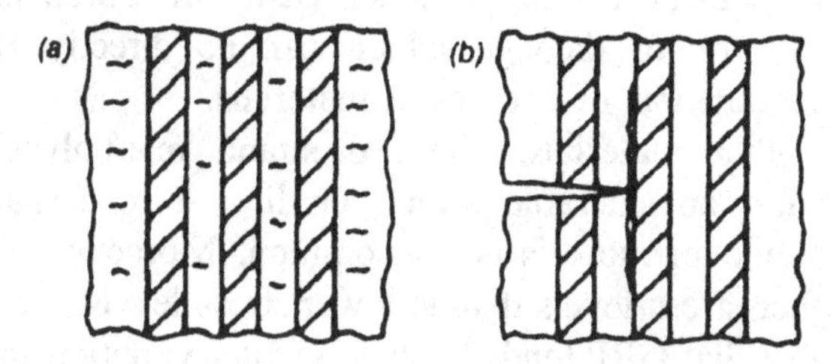

Ductile (polymer) matrix damage. a) Dispersed failure mode: cracks confined to matrix only. b) Localised failure mode: cracks grow by breaking fibres and, at later stage, interfaces fail.

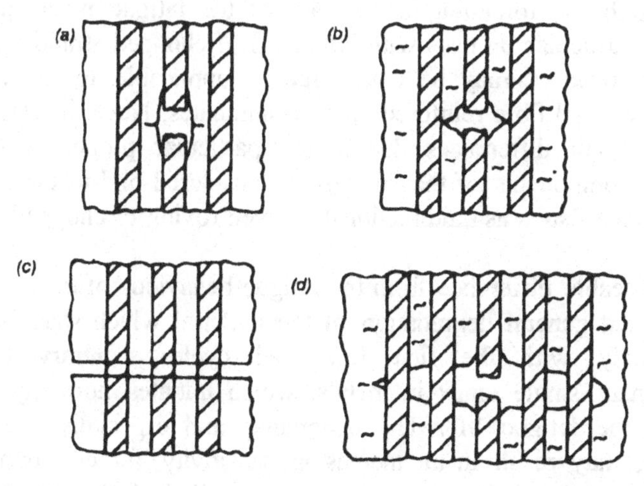

Fibre damage: a) fibre break causing interfacial debonding, b) fibre break increasing matrix crack, c) fibres bridging a matrix crack, and d) combination of a), b) and c).

Figure 10.3. Fatigue damage mechanisms in FRP [6].

The nominal debonding stress is cycle-dependent and so is the nominal resin cracking stress. It seems that transverse fibre debonding does not directly affect the strength, but when, under fatigue conditions the debonds propagate into resin cracks, some loss of strength can be expected.

The tests on CSM polyester laminate also showed a relationship between the fatigue damage and the modulus of elasticity of the laminate. Microscopic examination revealed that when the values of modulus had decreased by 2.5% debonding was well established. The loss in modulus at the onset of resin cracking was in the range of 8-10%.

In general, stiffness reduction is an acceptable failure criterion for many components which incorporate composite materials. In testing, stiffness change can be a rather precise, easily measured and easily interpreted indicator of damage which can be directly related to microscopic degradation of composite materials.

Like most other materials, chopped strand mat/polyester resin specimens exhibit considerable scatter on life. A scatterband of an order of magnitude or more is not uncommon. Moreover, the scatter is larger at higher stress levels than at lower stress levels, which might be due to the fact that GRP tends to show greater variation in ultimate tensile strength than metals. This is particularly so in the case of CSM laminates.

Although the foregoing description of the fatigue mechanism is based in particular on tests with laminates of chopped strand mat and polyester resin, it may be considered as applicable to the fatigue behaviour of all fibre reinforced plastic laminates. Nevertheless, there may be small differences due to the particular properties of the different components of the composite considered and to the type of reinforcement such as unidirectional, woven roving or chopped strand mat.

A noticeable difference from the fatigue behaviour of metals is the gradual and general degradation of the material which starts almost immediately with the first few load cycles, contrary to the predominant single crack in metals, which initiates normally about 30% of the fatigue life. The debonding and especially the resin cracking, may result in an increasing sensitivity for environmental influences. Debonding and cracking may result in leakage and, even worse, the penetrating media may be able to attack the fibre/resin-interface and thereby weaken the composite material causing premature failure.

So, bearing in mind the damage mechanism and the wide scatter, test results presented should preferably be as shown in figure 10.4 [9].

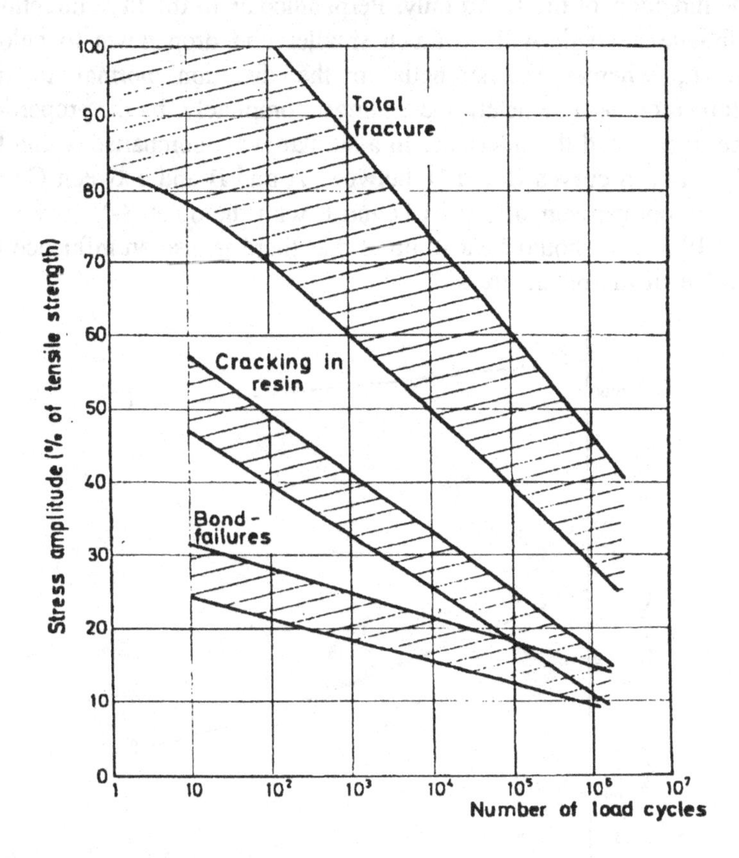

Figure 10.4. S-N curve for glass fibre polyester mats under alternating tension/compression. (R = -1, freq = 100 cpm).

10.4 FATIGUE TEST DATA

Typical S-N curves for a variety of GRP and CFRP laminates are given in figure 10.5 [8,10-13]. The figure shows clearly the large differences in fatigue strength which can be obtained, depending especially on the choice of the reinforcement and the ratio of fibre to matrix in volume or weight percentage. Comparing the curves A and C, the greater fatigue strength of curve A can be explained primarily by the greater strength of carbon fibre as compared with the strength of E-glass, secondly by the larger fibre-volume fraction and thirdly, to a smaller extent, because of better properties of epoxy resin as compared with polyester, among which is a better bonding strength. Both curves are related to the fatigue strength under tensile loading in the direction of the fibres only. Perpendicular to the fibre direction, the fatigue strength will be much smaller and drop down to below curve E, whereas the strength in the direction normal to the reinforcement will be determined, almost completely, by the properties of the matrix and the interface. In a similar way, comparisons can be made between curves B and D, between A and B and between C and D. In the comparison of curves D and E with the other S-N curves in figure 10.5, one should bear in mind that there is also an influence of the value of the mean stress.

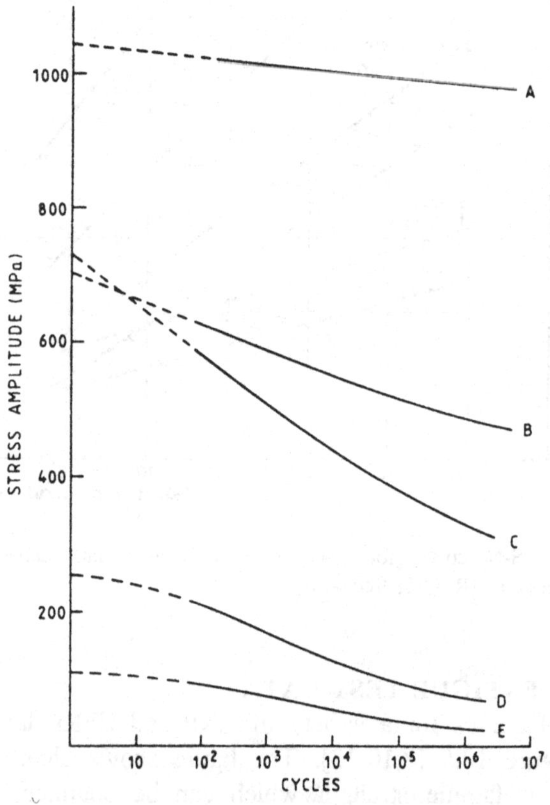

Figure 10.5. S-N curves for laminates with various types of reinforcement [8,10-13].
A: UD carbon fibre/epoxy (V_f = 0.6, tensile stress)
B: 0°/90° ± 45° carbon fibre/epoxy (V_f = 0.55, tensile)
C: UD E-glass/polyester (V_f = 0.4, tensile)
D: WR E-glass/polyester (V_f = 0.3, zero mean stress)
E: CSM E-glass/polyester (V_f = 0.2, zero mean)

With regard to the influence of the constituent materials on the fatigue strength of FRP, care has to be taken to distinguish between the influence of matrix material, reinforcement material and the construction and orientation of the reinforcement.

10.4.1 Influence of Resin Type

The resins most commonly used as matrices for GRP are polyesters and epoxy resins. The fatigue behaviour is similar to fatigue in metals and is cycle dependent rather than time dependent. Epoxide resins are slightly superior to phenolic, polyester and silicone resins. The better behaviour of epoxy resins is attributed to their greater strength, better bonding to the fibres, lower shrinkage resulting in smaller residual stresses and higher strain without cracking and exposure to fibre corrosion [3]. From cyclic flexural testing of polyester and vinyl ester resins it was concluded that vinyl ester type resins had a significantly better fatigue behaviour than polyester resin. Isophthalic polyester was somewhat better than orthophthalic polyester resin. Little difference was observed between the standard vinyl ester and the new pre-accelerated thixotropic vinyl esters [14,15]. Despite great chemical differences, the influence of the resin on the fatigue strength of FRP is rather small when compared with the influence of the different reinforcements.

10.4.2 Influence of Fibre Type

From tests by various investigators it is suggested that a ranking of materials from best to worst would be [14,16]:

- High modulus Carbon Fibre
- High strength and low modulus Carbon
- Aramid/Carbon hybrid
- Aramid
- Glass/Aramid hybrid
- S-glass
- E-glass

The better behaviour of carbon and aramid fibre reinforcement as compared to E-glass can be attributed partly to the higher ultimate strength as in the case of high strength carbon and aramid. However, the better behaviour under fatigue loading will be, above all, the result of the considerably greater Young's modulus, which results also in a much greater modulus of the laminate. As discussed before, the fatigue damage process starts with debonding of the fibres and cracking of the resin, both processes depending primarily on the strain cycling in the resin. Therefore a laminate with a greater modulus of elasticity needs a higher stress cycling level to reach the critical strain cycling level. This is especially the case under tensile loading conditions, as the resin is far less sensitive to cracking under compression. However, under compression load, the buckling strength of the fibre and shear forces in the interface between fibre and resin may come to dominate the fatigue behaviour, especially in the case of aramid but somewhat less with high modulus carbon fibres. This explanation seems to be in agreement with results of fatigue tests on carbon epoxy laminates [17].

10.4.3 Influence of Fabric Construction

Last but not least, the construction and orientation of the reinforcement plays a critical role in fatigue performance. In figure 10.6 various fabric constructions are compared with regard to their

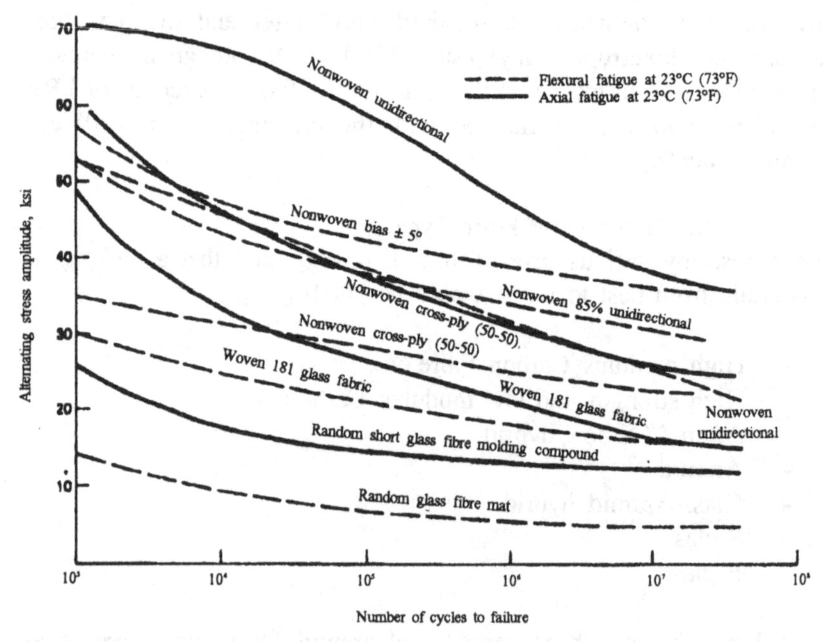

Figure 10.6. Comparative fatigue strength of various fabric constructions [14,18].

fatigue performance under alternating flexural and axial fatigue loading [14,18]. Again the rather high values of non-woven unidirectional reinforcement and the relatively low value of the random short glass fibre mat as compared to the strength of the woven roving can be explained largely by the differences in the fibre-volume fraction of the fibres orientated parallel to the loading direction. Other factors which influence the modulus of elasticity of the laminate and thus the fatigue behaviour will be the continuity of the fibres and the amount of stretching. Unidirectional fibres are continuous and rather well stretched, resulting in a rather high modulus of elasticity. The same applies for the non-woven fabrics. However, not all fibres are orientated parallel to the loading direction, resulting in a smaller load carrying fibre-volume fraction and thus in a lower modulus and a lower fatigue strength. In a woven fabric the load carrying fibre-volume fraction will be the same, but these fibres are not so well stretched, resulting in a decrease of modulus and of fatigue strength as compared with the unwoven reinforcement.

Finally, in the random short glass fibre mat, the load is no longer carried by continuous fibre, but the load force must be transferred from fibre to fibre by the matrix between the fibres. This results, on one hand, in a further reduction of the modulus of elasticity of the laminate and, on the other hand, in a higher stress level of the resins and especially in an increase of the shear forces in the interface between fibre and resin.

10.4.4. Other Influences

Further examination of figure 10.6 shows that for all fabrics the fatigue strength under flexural loading is smaller than under axial loading. The reason is probably the rather large stress gradient through the thickness of the laminate under flexural bending, which results in considerable shear stresses in the interface between fibres and resin.

In addition to the aforementioned material parameters, some other parameters may be discussed with respect to their influence on the fatigue strength.

On the material side, resin additions for one or another purpose, may effect the fatigue behaviour of the resin and of the laminate in an adverse sense. Whereas the fatigue damage process starts with the onset of debonding between fibres and resin, it seems clear that much attention has to be paid to the quality of the interface. Depending on the selected coupling agent a factor 4 to 5 in static bond strength of polyester to glass is possible [3]. It can be expected that this also will affect the fatigue life at lower stress levels.

The fatigue behaviour may also be affected by the production process and quality control. The curing and hardening process can affect the properties of the resin and the interface. During the laminating process air bubbles may be entrapped in the fabric weave or in the surface layer. Such voids may result in a fatigue strength reduction factor up to 1.3 or 1.4 for the onset of resin cracking [19]. The distribution of the resin, the saturation of the reinforcement and the stretching of the fibres are determined by the laminating process. Inhomogeneous distribution of the resin and slack of the fibres may result in an overall or local influence on stiffness and fatigue strength.

Fatigue life may be reduced considerably due to environmental conditions as in seawater, especially if the surrounding medium can penetrate into the laminate along surface cracks and along debonded fibres. The reduction of the fatigue strength is considerable at higher load amplitudes, whereas the effect will be small at lower stress levels [9,20,21]. In this respect it must be realised, that under a random load distribution resin cracking and penetration of the seawater may occur in a very early stage and may affect the fatigue strength during the following low load cycles more strongly than in the case of a constant amplitude loading. This shows the importance of a tough surface layer. Data on fatigue tests in seawater are restricted in number. Whereas tests in air are themselves limited in the applied frequency to about 5 Hz, this limitation is much more serious in seawater as the deterioration of the laminate by the environment needs time. Therefore the frequency of test in seawater should not exceed the frequency under working conditions or, at most, 0.2 Hz. A test up to 1 million cycles can be done in 2.5 days at 5 Hz, but a frequency of 0.2 Hz results in a test period of 2 months, which is expensive and time consuming.

By far the most fatigue data relate to tests under laboratory conditions on small specimens. Fatigue testing procedures are included in test standards (ASTM, BS); the requirements on test equipment and specimens are the same as for static testing. The specimens for tensile and bending strength properties are small, have a restricted width and the ends have a dog bone shape or bonded tabs. The scatter in the data to some extent reflects fluctuations in fabrication procedure and quality of workmanship, and to some extent the scatter may be caused by inconsistency and deficiencies in the underlying test methods. The small specimens are appropriate for 0°, 90° and randomly oriented reinforcing fabrics. Off-axis fibres (directions other than 0° or 90°) introduce stress concentrations into the laminate, which lowers the ultimate strength under tensile load. Stress concentrations also exist in the area of the taper or bonded taps.

In these cases the results of the testing may be conservative.

10.5 FATIGUE DESIGN AND DAMAGE RULES

Fatigue data in a form of S-N curves discussed so far do not show the influence of the mean stress on the fatigue behaviour. However, for design purposes this information may be essential, in which case more S-N curves for different mean stresses are needed. From these curves it will be possible to construct a Smith diagram or a master diagram in which the stress amplitude is given against mean stress for chosen lives. In Figure 10.7, such a master diagram is presented for an E-glass chopped strand mat polyester laminate [22]. When there have been insufficient test results to produce a master diagram, it has sometimes been suggested that an approximation of the diagram can be produced, using a linear relationship between the stress amplitude S_A, the mean stress S_M, the fatigue strength S_E at zero mean stress for a given life and the static ultimate strength S_u of the material:

$$S_A/S_E = 1 - S_M/S_u. \tag{10.4}$$

This relationship is known as the modified Goodman law and is applicable to many common metals. Regarding FRP, the use of S_c may be preferred in place of S_u. S_c is the stress rupture strength for the time corresponding to the cyclic endurance. However, investigations on the suitability of this law, or of alternative relationships for FRP, showed a restricted area of applicability and a rather poor accuracy of representation of material properties in fatigue [22].

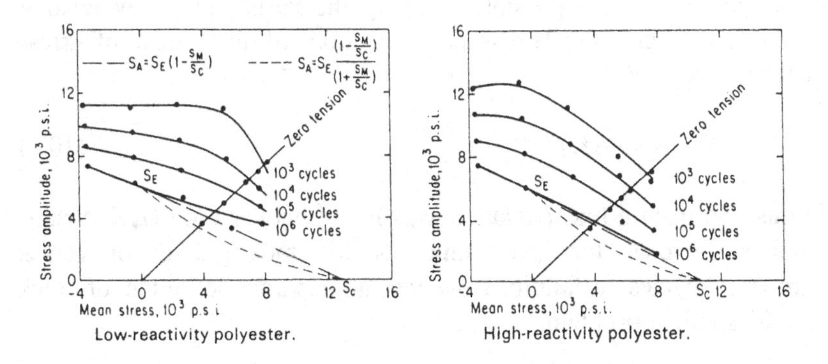

Figure 10.7. Master diagram for a chopped strand mat/polyester resin laminate [22].

An important aspect of safe-life design is the use of cumulative damage rules to predict fatigue life under conditions of varying or random loading from conventional constant stress amplitude fatigue data. Cumulative damage theories can be classified on the basis of the assumed relationship between damage and cycle ratio. The most widely used Palmgren-Miner rule is linear and hence stress-independent and free of stress interaction. This rule is commonly used for steel structures. Although the accumulation of fatigue damage in FRP is non-linear and stress-independent a linear law has been found to give reasonable results for FRP laminates subject to multi-stress level fatigue tests, i.e.:

$$\Delta = \Sigma_i n_i/N_i \tag{10.5}$$

where n_i is the current number of cycles at stress amplitudes, σ_i and N_i is the number of cycles to failure at σ_i. The sum is taken over all stress amplitudes and where Δ is a constant somewhat less than unity at failure. However, preference should be given to a non-linear damage law of the form:

$$\Delta = \Sigma_i\{A(n_i/N_i) - B(n_i/N_i)^2\} \tag{10.6}$$

where A and B are material constants and Δ is equal to unity at failure [7].

Cumulative damage expressions of the same form may also be related to initial fibre-debonding, resin cracking and to the reduction of the residual strength.

A start has been made with the fracture mechanics approach. It has been shown that it is possible to apply the Paris power law relating to the crack growth da/dN as a function of the range of stress intensity factor ΔK:

$$da/dN = C(\Delta K)^m. \tag{10.7}$$

In this approach a linear relation was found between log (ΔK) versus number-of-cycles for given amounts of crack growth or versus number-of-cycles to failure. However, an accurate definition of crack length is still a problem.

10.6 FATIGUE OF STRUCTURES

Fatigue data commonly refer to the laminate fatigue properties. However, in structures as a whole and in structural details, the risk of fatigue failure is larger at the locations of high stress concentrations or hot spots associated with notches, connections etc.

Fatigue testing on large scale specimens and service experience on GRP ships indicate that fatigue damage, comprising resin cracking and fibre-debonding at stress-raisers such as holes and hatch corners, usually remains very localised with negligible effect on overall structural behaviour [10,23].

However, fatigue is more likely to lead to serious problems at bonded structural connections, in which weakness is caused by the absence of load-bearing fibres across bonded interfaces and by the low interlaminar tensile and shear strength in combination with the inevitable occurrence of stress concentrations associated with joint geometry and bond imperfections. In many cases the occurrence of loads perpendicular to the plane of the laminate aggravates the problem. Purely theoretical estimates of joint strength seem unacceptable. Reference must therefore be made to test data, and development of a new high-performance design should include a thorough programme of tests on all important joints with evaluation of static and fatigue strength. Too little information is available at present.

Lastly, the behaviour of sandwich structures under fatigue loading will be considered. In a sandwich structure different materials are applied for skin laminates and core, fixed together by some form of adhesive bonding and subjected to different types of load. Commonly bending load will dominate in sandwich panels, resulting in tensile or compressive stresses in the skin and in shear forces in core and in the adhesive bonding. So, in addition to fatigue data of FRP laminates, data for the fatigue strength of the core materials are also necessary.

Sandwich structures have been tested in four-point bending under normal fatigue loading as well as under slamming fatigue, see figure 10.8, [24]. For the core material of the specimens linear and rigid, cross-linked PVC foams have been applied. During the tests a continuous but small increase of deflection could be observed. However, the stiffness of the sandwich remained constant during most of the fatigue life. Only shortly before failure could changes in stiffness be observed, while also large shear deformations in the core became visible. If the fatigue data are normalised by the quasi-static yield strength of the core, all data could be presented by one line in a linear-log diagram. Both fatigue strength and static strength under shear forces are related to the density of the foam.

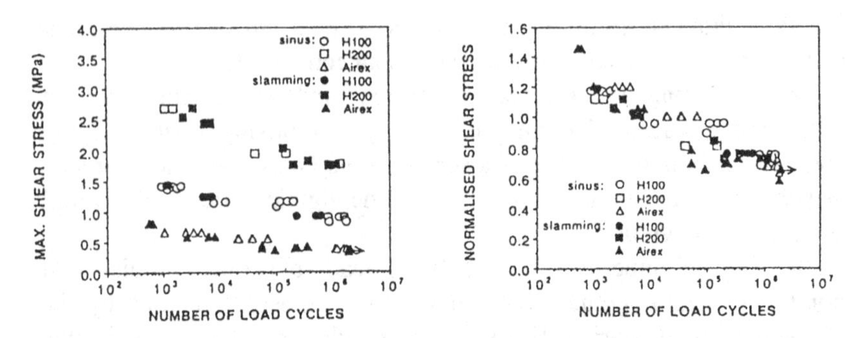

Figure 10.8. S-N curves for sandwich beams under sinusoidal and slamming fatigue [24].

10.7 CONCLUSIONS

For design purposes, many theories and methods arc developed to describe the fatigue strength of composite materials. However, given on the one hand the broad range of applications and the variety of composites applied in marine structures and on the other hand the limitations of most theories, theoretical calculations as to the fatigue life of a given composite should only be used as a first-order indicator. The best method of determining the fatigue properties of a candidate laminate or structural detail is fatigue testing in an experimental programme for obtaining additional information. Although precise prediction of fatigue life for FRP laminates is currently beyond the state-of-the-art of analytical techniques, some insight into the relative performance of constituent materials can be gained from published test data.

In the use of test data utmost care must be given to the particulars of specimens and test-conditions and to all influencing parameters in production and in testing of the specimens.

Sound comparisons and predictions can be made only if all the parameters influencing the test data are well-known.

10.8 REFERENCES

1] Almar-Naess, A., (ed.), "Fatigue Handbook - Offshore Steel Structures", Tapir Publishers, Trondheim, Norway, 1985. ISBN 82-519-0662-8.

2] Maddox, S.J., "Fatigue Strength of Welded Structures", Abington Publishing, Cambridge, England, 1991. ISBN 1-85573-013-8.

3] Dew-Hughes, D., Way, J.L., "Fatigue of Fibre-Reinforced Plastics: A Review", Composites, July 1973. pp 167-173.

4] Harris, B., "Fatigue and accumulation of damage in reinforced plastics", Composites, October 1977. pp 214-220.

5] Reifsnider, K., "Fatigue Behaviour of Composite Materials", Int. J. Fracture, **16**, (1980), pp 563-583.

6] Talreja, R., "Fatigue of Composite Materials", Technomic Publishing: Lancaster, Pennsylvania 17604, USA, 1987. ISBN 87762-516-6.

7] Owen, M.J., Howe, R.J., "The Accumulation of Damage in a Glass-Reinforced Plastic under Tensile and Fatigue Loading", J. Phys. D.: Appl. Phys., **5**, 1972. pp 1637-1649.

8] Owen, M.J., Smith, T.R., Dukes, R., "Fatigue of Glass-Reinforced Plastics, with Special Reference to Fatigue", Plastics & Polymers, June 1969. pp 227-233.

9] Malmo, J., "Fatigue Properties of Glass Fibre Reinforced Polyester", Proc. Symp. *Use of Reinforced Plastics in the Petroleum Industry*, The French Petroleum Institute, Paris, 15th-16th June 1978.

10] Smith, C.S., "Design of Marine Structures in Composite Materials", Elsevier Science Publishers Ltd., London, 1990. ISBN 1 85166 416 5.

11] Agarwal, B.D., Broutman, L.J., "Analysis and Performance of Fiber Composites", Wiley, New York, 1980.

12] Boller, K.H., "Fatigue Characteristics of RP Laminates Subjected to Axial Loading", Modern Plastics, **41**, 1964. p 145.

13] Howe, R.J., Owen, M.J., "Cumulative Damage in Chopped Strand Mat/Polyester Resin Laminates", Proc. 8th Intl. Reinf. Pl. Congr. BPF, London, 1972.

14] Greene, E., "Marine Composites: Investigations of Fiberglass Reinforced Plastics in Marine Structures", Ship Struct. Cttee., USCG, Rep. SSC-360, September 1990.

15] Burrel, et al., "Cycle Test Evaluation of Various Polyester Types and a Mathematical Model for Projecting Flexural Fatigue Endurance", Proc. 41st Annual Conf. Society of the Plastics Industry, 1986.

16] Konur, O., Mathews, L., "Effect of the Properties of the Constituents on the Fatigue Performance of Composites: A Review", Composites, **20**, (4), July 1989. pp 317-328.

17] Schütz, D., Gerharz, J.J., "Fatigue Strength of a Fibre-Reinforced Material", Composites, **8**, (5), October 1977. pp 245-250.

18] "Engineers' Guide to Composite Materials", Am. Soc. Metals,

Metals Park, Ohio, 1987.

19] Owen, M.J., Griffith, J.R., "Evaluation of Biaxial Stress Failure Surfaces for a Glass Fabric Reinforced Polyester Resin under Static and Fatigue Loading", J. Mater. Sci., **13**, 1978. pp 1521-1537.

20] McGarry, et al., "Marine Environment Effects on Fatigue Crack Propagations in GRP Laminates for Hull Construction", Report No. MITSG 73-16, MIT, Boston, 1973.

21] Dixon, R.H., Ramsey, B.W., Usher, D.J., "Design and Build of the Hull of HMS Wilton", Proc. Intl. Symp. *GRP Ship Construction*, RINA, London, October 1972.

22] Smith, T.R., Owen, M.J., "Fatigue Properties of RP", Mod. Pl., April 1969. pp 124-132.

23] Report of Cttee. III.2, *Non-ferrous and Composite Structures",* Proc. 9th Intl. Ship Struct. Cong., Santa Margherita, September 1985.

24] Bergan, P.G., Buene, L., Ecktermeyer, A., "Assessment of FRP Sandwich Structures for Marine Applications", Proc. Charles Smith Memorial Conf. *Recent Developments in Structural Research*, DRA, Dunfermline, Scotland, July 1992.

11 COMPOSITES IN OFFSHORE STRUCTURES

11.1 BACKGROUND

Composites offer the possibility of significant savings, in platform topside weight, as well as in installation and through-life maintenance costs when used offshore. FRP has been used to a limited extent in a range of applications on offshore platforms and, despite some of the problems mentioned below, there is now increasing interest in its wider use.

Use of composites in load-bearing elements, both offshore and in other structural applications, is being hindered to a certain extent by a lack of design codes. Recently, a number of joint industry projects have been initiated in an attempt to rectify the situation and the recommendations of these will become available in due course.

One particular obstacle to the use of composites has been their perceived flammability. Several testing programmes have been carried out to quantify fire performance and it has been found that, although the resin phase is indeed potentially combustible, the composite itself often shows interesting and potentially useful effects in fire. The most significant of these effects is the slow rate of burn-through and heat transmission that can be achieved under certain conditions. This property is already being exploited in some piping and panel applications.

There are some limitations connected with the fabrication technology needed to make large structures in composites. The scale of many components in use offshore is larger than in other engineering fields and, unlike the fabrication technology for metals, forming processes for composites are not yet so adaptable to large scale construction. Although there are many processes which are used to fabricate composites, only four of them: contact moulding, resin transfer moulding, pultrusion and filament winding, have the potential for the efficient production of large components and structures.

In recent years a number of research programmes [1-7] and studies have been carried out with the aim of removing obstacles mentioned above. The potential now exists for FRP to enter several areas: firewater piping and other aqueous services, walkways and flooring, structural and semi-structural walls and floors, especially where blast and fire protection is required.

Some existing offshore applications of FRP are listed in table 11.1. In many of these it can be shown that, taking into account all relevant lifetime factors, composite designs can fulfil the necessary overall performance requirements. In many cases advantages in favour of the composite solution can be identified, either in terms of weight-saving, safety or reliability improvement.

Table 11.1. Existing applications of composites offshore.

- Fire protection panels	- Pipe refurbishment
- Water piping systems	- Corrosion protection
- Walkways and flooring	- J-tubes
- Partition walls	- Casings
- Tanks and vessels	- Lifeboats
- Cable ladders and trays	- Buoys and floats
- Boxes, housings and shelters	- ESDV protection

In the future, a wider range of load-bearing applications may be possible. Applications already under consideration [8] include high pressure pipework, tethers, risers, tubing, core sample tubes and drill pipes. GRP is also being considered for the corrosion and fire protection of jacket members. When the fabrication technology is more advanced than at present, GRP may eventually become a competitive material for larger structural elements.

11.2 SELECTION OF MATERIALS

One of the first problems faced by designers unfamiliar with composites is that of making the choice of the appropriate material from the array of different reinforcement and resin combinations.

11.2.1 Reinforcements

For offshore installations the choice of reinforcement is simplified, since cost constraints render the more expensive high performance reinforcements, carbon and aramid, unattractive. The emphasis for tonnage use is strongly on glass fibre, which can be employed in a variety of forms, including unidirectional tows, woven fabrics and random mats. There are, of course, some additional areas where the special properties of the high performance fibres render their use cost-effective (combinations of aramid and carbon fibres with S-glass, for instance, are being considered for use in risers [8]), but glass is the main reinforcement type under consideration at present.

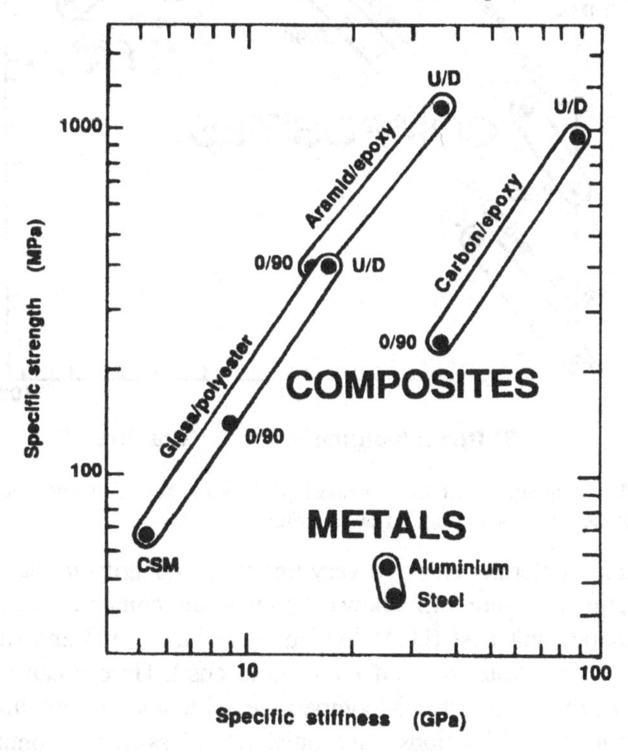

Figure 11.1. A comparison of the relationship between specific strength (UTS/specific gravity) and specific stiffness (Young's modulus/specific gravity) for a range of engineering materials.

Some key property parameters in the selection of structural materials for offshore use are demonstrated in figures 11.1 and 11.2. Engineering materials for weight-critical applications are frequently compared on the basis of specific strength (strength per unit weight, expressed as the tensile UTS divided by the specific gravity) and specific stiffness (stiffness per unit weight: modulus divided by specific gravity) [9,10]. Figure 11.1 compares various types of fibre reinforced plastic, structural steel and aluminium in terms of these quantities. While all the composites shown offer specific strength advantages over metals, it can be seen that only the 'high performance' composites, which are the most expensive, can outperform the metals in terms of specific stiffness.

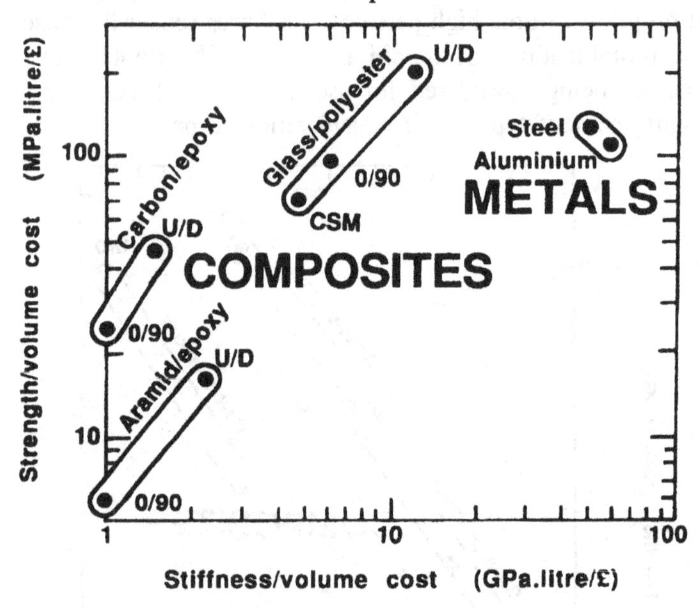

Figure 11.2. Comparison of the relationship between strength/volume cost and stiffness/volume cost for a range of engineering materials.

Of course, materials cost is a very important quantity in the case of large structures. Figure 11.2 shows the materials comparison in terms of strength per unit cost (UTS divided by volume cost) and stiffness per unit cost (modulus divided by volume cost). Here it can be seen that none of the composites is competitive with steel or aluminium in stiffness-critical applications and only the glass-based composites compete in strength-critical areas. This is the reason why, for tonnage structural and semi-structural usage, glass rather than carbon or aramid-based composites are the materials of main interest. Most of the research programmes relating to the offshore industry have, for this reason, concentrated on GRP rather than the 'advanced' composites. It should be noted that the comparison in figure 11.2 is made on the basis of the raw material costs. If installed component costs (or perhaps even through-life costs) had been used instead, the results would have shown a much clearer advantage for GRP.

As can be seen from figures 11.1 and 11.2, it is possible, with composites, to tailor the mechanical properties to suit the final application by controlling the orientation and disposition of the reinforcing fibres. As discussed in previous Chapters, glass fibre can be made available for processing in a number of forms, including unidirectional rovings, woven fabrics (made from unidirectional rovings) and random mats, which can contain either discontinuous fibres (chopped strand mat) or continuous fibres (swirl mat). The fibre-volume fraction in the composite is largely determined by the format of reinforcement chosen, because the extent to which fibre content can be varied by changing the processing conditions is quite limited. High fibre-volume fractions are desirable for use offshore, in order to maximise both mechanical and fire performance. Unidirectional rovings (as employed, for instance, in filament wound pipework and in pultrusions) give the highest fibre-volume fraction, usually in the range 0.5-0.65. In woven fabrics the volume fraction is generally 0.4-0.55 and with random mats, which give the lowest fibre content, 0.25-0.33 is achievable.

11.2.2 Matrix Resins

The selection of resin matrices is important, since the matrix plays a critical role in determining off-axis strength, damage tolerance, corrosion resistance and thermal stability. Fabrication technology restricts the field to thermosetting resins at present but there are still five candidates, as shown in table 11.2, each with particular advantages and drawbacks.

Table 11.2. Candidate resins for use in composites for the offshore industry.

	Cost (£/tonne)	Mechanical Strength	Corrosion Resistance	Fire Performance
Polyester	1200-1600	xx	xx	x
Vinyl ester	2200-2600	xxx	xxx	x
Modar	2000-3000	xx	xx	xxx
Epoxy	> 4000	xxxxx	xxxxx	x
Phenolic	1300-1700	xx	xx	xxxxx

11.2.2.1 Unsaturated polyesters

Unsaturated polyesters are the resins most widely used in GRP. Their principal advantage, besides low cost, lies in their cure chemistry. The free radical cure reaction, triggered by an addition of a peroxide initiator, offers a rapid but controllable cure, while the resins themselves have a long shelf life. For this reason, polyesters are very easily fabricated. There are several types of polyester but isophthalic resins offer the most attractive combination of mechanical strength and resistance to the marine environment. Moreover, isophthalics are already widely used in marine applications, most notably in the hulls and superstructures of minehunters.

The disadvantage of polyesters is their fire performance, which as will be demonstrated later, is relatively poor in terms of toxic product and smoke production. This may limit their application in sensitive areas such as accommodation modules. However, the retention of integrity of polyester laminates in fire is good, as will be shown later, and their flammability characteristics can be modified to some extent by the use of additives, although this can increase the toxicity of combustion products. One of the most attractive non-toxic fire retardant additives, alumina trihydrate (ATH) works well in polyesters, but raises the resin viscosity, thus making fabrication more difficult. There is a trend, wherever possible, away from fire retardants that produce toxic products. The resins used in minehunters, for instance, are not fire-retarded.

11.2.2.2 Vinyl esters

Vinyl esters lie mid-way in properties between polyesters and epoxies. While retaining some of the ease of fabricability of the free radical cure they offer better mechanical properties and are often preferred in demanding applications, particularly those where chemical or environmental resistance is needed. Applications of vinyl ester relevant to offshore use, include pultruded gratings for walkways, as well as pipes and tanks.

The most significant recent application of vinyl esters, involving large diameter filament wound pipework has been in the oil and ballast water systems of the gravity base platform, Draugen [11]. This pipework is designed to cope with both water and oil at temperatures of 40°C, with possible short excursions up to 70°C.

11.2.2.3 Modified acrylics (Modar)

Urethane methacrylate or 'Modar' resins cure very rapidly and are often favoured for processing reasons for use in pultrusion. The low viscosity of the base resin permits the incorporation of high levels of the dire retardant, ATH. Pultruded Modar products can be made with fire performance which is significantly better than that of the resins mentioned previously, which has led to uses of the material in fire-critical applications such as public transit systems and for the cable ducting in the Channel tunnel. The Modar resins, unlike the other candidates discussed here, cannot be processed by open mould methods and must therefore by fabricated by pultrusion or resin transfer moulding.

11.2.2.4 Epoxy

Epoxy resins, of which there are many variants, have perhaps the most outstanding combination of strength, toughness and corrosion resistance of the resins commonly used in composites. They are, however, expensive and fabrication can be a little more difficult than with the free radical cured thermosets mentioned previously. The most significant current application of epoxies in relation to offshore use, is in the manufacture of filament wound pipework.

11.2.2.5 Phenolic

Phenolic resins, the oldest class of synthetic polymer, have outstanding thermal and fire performance, as shown for instance by their traditional uses as binders for foundry sands and brake linings. Unlike many other thermosets, which depolymerise or decompose under fire conditions to give undesirable gaseous products, phenolics, which contain a high percentage of aromatic material, undergo progressive condensation of the aromatic rings to form an intractable char, which protects the surface of the composite. Phenolics have low initial flammability and, when involved in a fire, they contribute little further heat, producing only low levels of smoke and toxic products [12].

Difficulties with phenolics have traditionally been associated with cure chemistry, poor control of the cross-linking reaction and the fact that water is evolved as a condensation product during cure. However, recent developments in cure chemistry have resulted in resins which are much easier to process than previously.

The water evolved during cure leads to laminates which have a distribution of microvoids and has raised some doubts about mechanical properties. Recent work with the new generation of phenolic laminates, however, has shown that, although the matrix resin is brittle, the mechanical strength and durability of the composite are comparable with those of glass/polyester composites.

The void content does, however, lead to higher levels of water ingress than with other types of laminate, and water content can be a problem in fires, because steam evolution can result in premature delamination. This problem can be minimised by the use of special through-stitched reinforcing fabrics which have improved resistance to delamination.

Although phenolic composites are difficult to process by pultrusion, developments are taking place in this area and pultruded phenolics may become available in the future. Phenolics are sometimes the only composites permitted in certain fire-critical applications. They are certainly strong candidates for use in panels for external cladding and for accommodation areas.

11.3 MANUFACTURING PROCESSES AND PRODUCTS

Although the range of composite fabrication processes is wide, offshore applications are unusual in that the potential size of components and structures is much larger than that encountered in most other areas. Much of the technology is geared to the manufacture of shell-like structures where the thickness is of the order of a few millimetres. Many potential applications on offshore platforms involve laminates of significantly greater thickness, or assemblies of composite elements. There are, therefore, only a few processes which are adaptable to the required scale. As mentioned in section 11.1, these include:

- contact moulding (for panels and other shapes)
- resin transfer moulding (for panels and other shapes)
- pultrusion (for sectional members including structural items, cable trays, planks and also skins for panels)
- filament winding (for tubes, risers, storage tanks and vessels)

The near future will probably bring advances in the scale, reliability and economic effectiveness of these processes, leading to the capability for construction of increasingly larger artifacts in a cost-effective manner.

11.3.1 Manufacture of Panels and Plates

The simplest technique for fabricating thick, large area panels and plates is the semi-automated process employed for the production of internal structures and superstructures for ships. In this process, as used for instance by Vosper Thornycroft, flat composite panels and plates which can be rib-stiffened or of sandwich construction, are laid-up and fabricated into modules of a size which can be readily handled and assembled into larger units. It is interesting to note that such plate structures are designed and loaded in a manner not dissimilar to that encountered with steel plates, with the exception that extensive use is made of adhesive bonding technology, rather than welding, in final assembly. This form of construction is adaptable in principle to many of the modules used on platforms, including accommodation areas.

In late 1990, 30 tonnes of fire protection panels, containing polyester resin, were manufactured by Vosper Thornycroft, and supplied to Amerada Hess for use of the helideck and part of the accommodation area of the Ivanhoe/Rob Roy rig. Installation of these panels is shown under way in figure 11.3.

Figure 11.3. Installation of GRP heat protection panels on the helideck of the Amerada Hess Ivanhoe/Rob Roy platform (Courtesy of Vosper Thornycroft Ltd).

11.3.2 Pultrusion of Sections and Panels

Pultruded sections, in glass/polyester, glass/vinyl ester and glass/Modar are beginning to be used as gratings, walkways and decking, as shown in figure 11.4. In an early example of this type of application, a composite well bay area, consisting largely of pultruded decking elements, was installed in 1986 on Shell's Southpass 62 production platform in the Gulf of Mexico. This replaced a heavily corroded steel structure. As mentioned previously, pultruded sections are also used as cable trays and electrical conduits.

Figure 11.4. Installation of Duradek pultruded grating on the mezzanine deck of the Shell Oil 'Ellen Rig' Beta unit (Courtesy of Fibreforce Composites Ltd).

The process of pultrusion [13] is not limited to unidirectional reinforcements: mat-type reinforcements can also be used. The properties of pultruded sections can therefore be tailored to particular applications [14]. A typical pultruded section may often contain unidirectional reinforcement at the centre, with skins of random mat reinforcement. As shown in figure 11.5, there is considerable scope for tailoring the flexural properties of the section by varying the proportions of unidirectional and random reinforcement [15].

One early area of concern in relation to pultruded gratings of the type shown in figure 11.4 was impact behaviour in the case of dropped object and similar types of loading. This problem was addressed in the Marinetech North West Phase I programme [6]. It was found that, while individual composite testpieces showed brittle behaviour in simple tests, composite structures such as gratings, are capable of absorbing significant amounts of energy by a mechanism of progressive damage to the structure. It was also found [6] that impact energy absorption increased with loading rate, certainly up to the loading rates experienced in dropped object impact. Conservative designs for impact resistance can, therefore, be realised on the basis of quasi-static tests on the type of structure concerned.

Pultrusion is not only viable for compact sectional structures. The

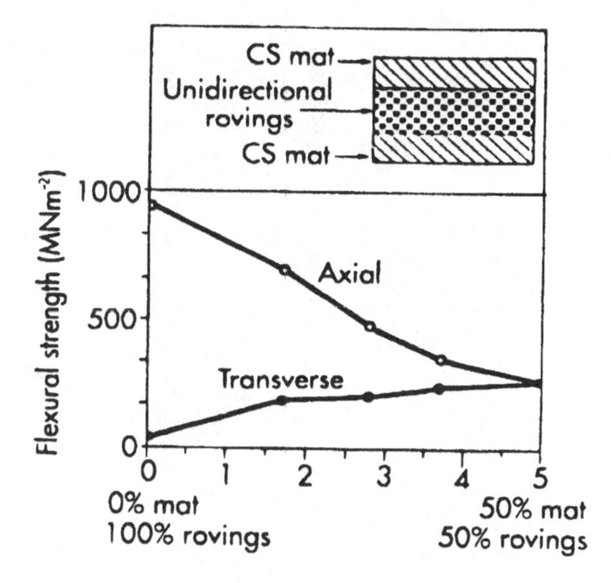

Figure 11.5. Effect on flexural strength of varying the ratio of continuous strand mat to unidirectional rovings in pultruded glass/polyester composites [15].

process may also be used for panels and plank-like components, the current width limit being about 1 m.

11.3.3 FRP Piping

The filament winding process, where unidirectional tows are wrapped around a core or mandrel at a prescribed angle, is a flexible process which can be used to make pipes and a wide variety of other products. The process is amenable to scale-up, and tanks and storage silos for the chemical industry are already being made with diameters of several metres. The primary area of interest offshore is in firewater piping (for pressures up to about 10 bar) and other aqueous services. The most widely used method of pipe assembly currently involves the use of adhesively bonded joints.

Several companies are undertaking development on the use of GRP pipes and vessels, the main emphasis being on filament wound glass/epoxy.

Experience with GRP pipes is being gained on several platforms, mainly in the Gulf of Mexico and off the coast of Africa. Figure 11.6 shows part of the firewater system installed by Elf on the Tchibouela platform in the Congo.

At present, use is confined mainly to aqueous systems at relatively low pressure, but more stringent applications are imminent. In early

Figure 11.6. Firewater system manifold on the Elf Tchibouela platform in the Congo (Courtesy of Elf Aquitaine).

work relevant to offshore use, successful shipboard trials were performed on filament wound pipes on oil tankers, both for water ballast piping [16] and as exposed weather deck piping system with all-GRP supports [17]. GRP performed well in both types of application. Moreover, this type of piping was shown to have good fire performance when filled with water [18,19,3]. In some circumstances of course, firewater piping may be required to be dry prior to use, in which case it has been shown that the required performance can be achieved by the application of various types of intumescent of cementitious coating. The regulatory position regarding the use of GRP pipes in ships is a little ahead of that in the offshore industry: draft IMO regulations for the use of GRP pipework have already been drawn up [19].

With increasing experience, and as a result of technology transfer from the chemical industry, where GRP piping is already widely used, further applications can be expected which will involve process fluids as well as aqueous systems. The capabilities of resin systems such as vinyl ester, which are already recognised elsewhere, are now beginning to be recognised as shown by the choice of these resins for the large diameter pipework in the base of the Shell Draugen facility [11].

Many of the North Sea applications of composites have been in retrofit areas. The approval procedure appears, as a matter of practicality, to be more straightforward in this type of application. The classic case in the Norwegian sector, which will no doubt form the basis of many future retrofit applications of firewater piping, has been the Valhall project.

This involved four oil companies, Amoco, Conoco, Shell and Statoil, and considered the replacement of a conventional metallic firewater system, with FRP [20]. The deluge system was of the partially dry type, where key elements were expected to be empty when not in use. After detailed considerations of possible fire occurrences it was concluded that, in the worst likely case, the dry part of the system would be exposed to a blast, followed by a hydrocarbon fire for a period of five minutes before being filled with water at 10 bar, at which time the active deluge system reduced the temperature to a value below 650°C. The test programme involved subjecting a sample pipework system to a mechanical loading equivalent to a blast, followed by hydrocarbon jet and pool fires. The work showed that GRP pipework with added passive fire protection performed more than adequately and in fact was likely to provide a safer, more reliable solution than a metallic system.

It is to be hoped that, driven by demonstrable improvements, safety regulatory requirements may be changed at some time in the future to permit the widespread use of GRP firewater systems in the North Sea.

11.4 FIRE RESISTANCE OF COMPOSITE STRUCTURES

Although the organic matrix material is intrinsically combustible it became apparent sometime ago that polymer-based composites, especially in thick sections, possessed desirable properties in fire. It was observed, for instance, during early experience of the use of GRP in minehunters, that the low thermal conductivity of these materials (200 times less than that of steel) was advantageous in the prevention of fire spreading from one location to another. Thick section laminates, with the higher glass contents achievable by the use of woven glass-reinforcements can show remarkable structural integrity in fire. One of the earliest exploitations of this property was probably by Vosper Thornycroft in the heat protection panels shown in figure 11.3.

A major aim of Phase I of the Marinetech programme [6], in which the author has been involved, was to characterise the fire performance of composites based on the candidate resin systems. The four resins included, which were the ones capable of being processed by contact moulding, were: polyester, vinyl ester, epoxy and phenolic. Source details are given in table 11.3. Wide-ranging comparative tests were carried out in order to provide a basis for choice between the different resin systems. A secondary aim was to begin to quantify the beneficial laminate thickness effect. Two of the techniques employed, and a few of the results obtained, will be briefly discussed here.

Table 11.3. Materials evaluated in the Marinetech North West Fire Test Programme.

Reinforcement	OCF Fiberglass woven rovings (1 ply ~ 0.5 mm)
Resins (all contact moulding grades)	
Isophthalic polyester	DSM Stypol 73/2785
Vinyl ester	DSM Atlac 580/05
Epoxy	Ciba Geigy Araldite LY1927/HY1927 (amine cure)
Phenolic	BP Cellobond J2018/L + Phencat 10 hardener

11.4.1 Characterisation of Composite Fire Performance

11.4.1.1 NBS cone calorimeter

One of the most important factors in fire has been identified as heat release rate which, in certain circumstances, can be critical in determining both the intensity of a fire and the rate at which it becomes fully developed. Until recently, calorimetric methods of measuring heat release were inaccurate and subject to significant experimental variability. A recently developed instrument, the cone calorimeter, shown schematically in figure 11.7, has significantly changed this situation. Cone calorimetry employs an indirect method of measuring heat release. The application of the technique to the characterisation of resins for use offshore has been outlined recently by Hume [21].

Cone calorimetry relies on an empirical law, the 'oxygen

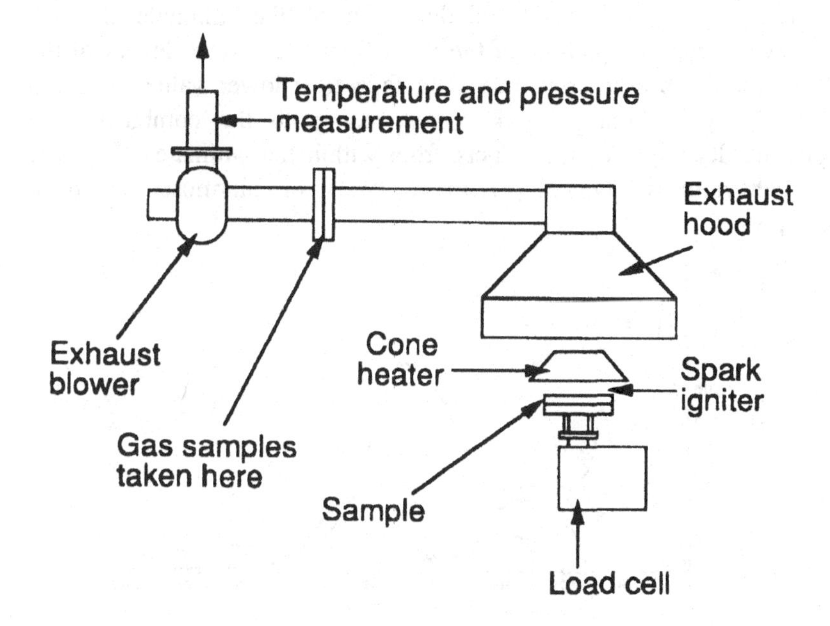

Figure 11.7. Schematic diagram of the NBS cone calorimeter.

consumption principle', which recognises that the ratio of heat released in a fire to oxygen consumed is, to a close approximation, constant for most materials. Measuring the oxygen concentration in the exit gas stream of the apparatus, therefore, enables the heat release rate to be inferred indirectly, but more accurately than in previous types of direct measurement.

The power supply to the ceramic cone of the calorimeter can be adjusted to give an irradiance characteristic of a particular type of fire exposure. In addition to oxygen concentration, it is also possible to measure the concentration of toxic gas species such as carbon monoxide and dioxide in the exit stream, as well as smoke density. Experimental details of the cone calorimeter technique are given in ISO DIS 5660. However, one modification of the procedure for composites testing is the need for specimen edge protection (using a 'picture frame') to avoid errors duc to the early burning of gaseous products from the cut edges of the laminate.

Figure 11.8 shows a comparison of typical heat release results obtained from laminates containing polyester, epoxy and phenolic resin. Vinyl ester has been omitted as it shows very similar behaviour to polyester resin. In the case of the polyester and epoxy there is a small induction period followed by a rapid rate of heat release, corresponding to burning of the resin at the laminate surface. Following rapid depletion of the resin from the surface layers of the laminate the heat release rate then falls to a lower value, before a second, much broader peak corresponding to the combustion of gaseous decomposition products from within the laminate. This type of behaviour is typical of many types of thermosetting resin composite.

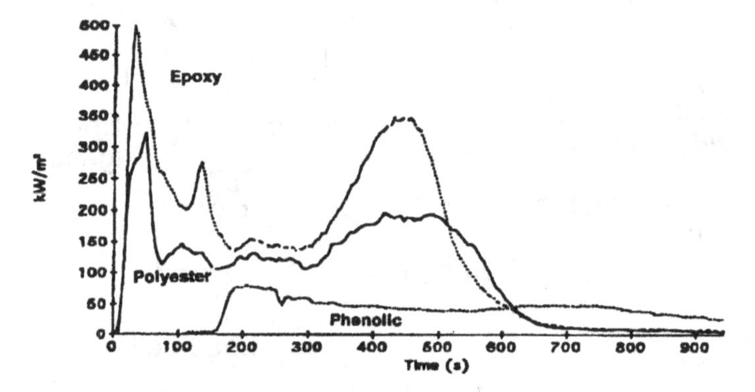

Figure 11.8. Comparison of heat release rate versus time for three resin systems, polyester, epoxy and phenolic at an irradiance of 80 kW/m^2. Woven roving laminates. Thickness: 6 plies (approximately 3 mm).

In contrast to the polyester and epoxy systems, the phenolic resin laminate shows behaviour of a very different type; the time to ignition is longer than with the other resins and both the peak heat release rate and the overall heat release are lower.

Figure 11.9 shows the peak heat release rate and the smoke density as a function of irradiance for the resin systems in question. Once again the behaviour of the phenolic resin laminates is substantially different from that of the others. It should be noted that the irradiance levels in real hydrocarbon fires, which have been the subject of some debate, are higher than those achievable in the cone calorimeter, possibly of the order of 200 kW/sq.m. However it can be seen that the heat release begins to approach a steady value with increasing irradiance.

Figure 11.10 shows the results of an investigation of the effect of laminate thickness, up to ~21 mm, for four resin systems with woven glass reinforcement. It can be seen that as the laminate thickness increases the peak heat release rate falls to a value that is much lower than that normally observed for thin laminates. To take advantage of the thickness effect it can be seen to be desirable to have laminates of the order of 8 mm or more in thickness.

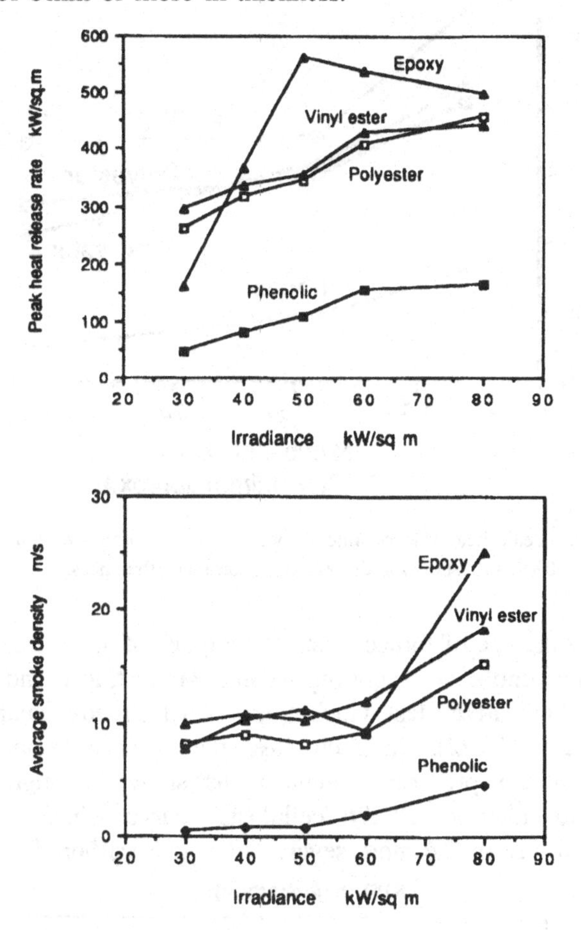

Figure 11.9. Peak heat release rate and average smoke density versus irradiance for four resins systems, polyester, vinyl ester, epoxy and phenolic. Woven roving laminates. Thickness: 6 plies (approximately 3 mm).

In summary, the cone calorimeter results show that the phenolic resin system is superior to the other resins in terms of both heat release and smoke density. It also indicates that there is a laminate thickness effect which may possibly be used to advantage in the design of fire-resistant structures.

11.4.1.2 Furnace tests

The fire barrier performance of conventional materials is generally

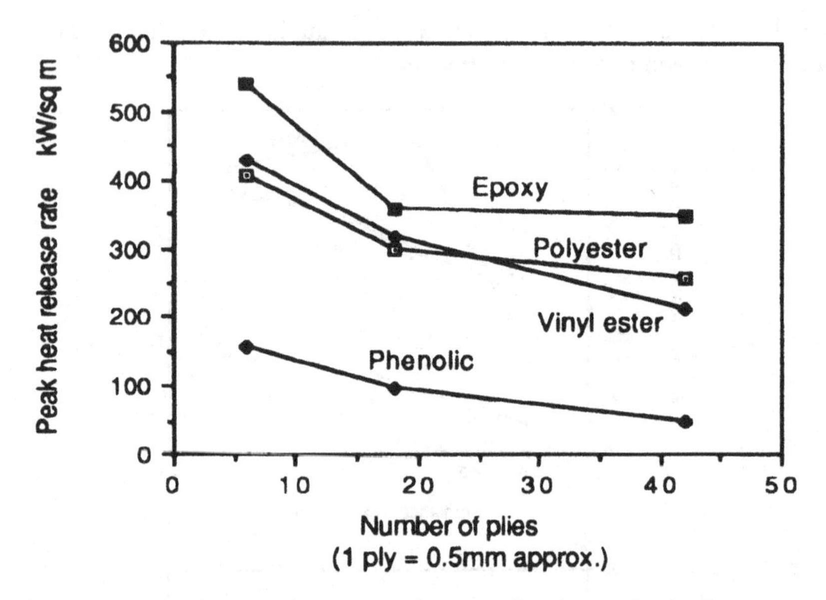

Figure 11.10. Peak heat release rate of woven glass composite laminates at an irradiance of $60\,kW/m^2$, showing the effect of laminate thickness.

assessed by the use of furnace tests. Two types of test are carried out: a large scale simulation, requiring a 3 m x 3 m testpiece and the more convenient 'indicative' test which employs a 1.2 metre square sample (BS476, parts 20-22). In each case the furnace temperature is increased in the prescribed manner, as shown in figure 11.11, according to either the BS476 'cellulosic' curve, which simulates a conventional fire or the more severe NPD hydrocarbon fire curve.

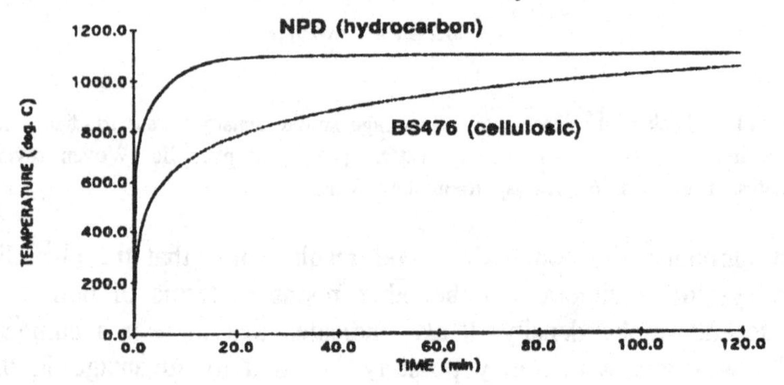

Figure 11.11. Fire test curves, according to the NPD (hydrocarbon) and BS476 (cellulosic) specifications.

The retention of integrity of thick laminates in fire is impressive, as can be seen from the results in table 11.4. All the laminates require a period of the order of hours for complete burn-through to occur under the conditions of the cellulosic test curve. For the more stringent NPD hydrocarbon curve the burn-through times are greatly reduced, although the phenolic resin composite still lasts for more than an hour. Burn-through of the laminate occurs by a progressive process in which the resin is depleted from the surface layers, leaving successive plies of the glass reinforcement, which eventually fall away.

Table 11.4. Fire resistance of 9mm thick woven roving laminates in the indicative fire test.

Resin Type	Time to penetration	Time to 160°C
Polyester		
Cellulosic curve	182 min	28 min
NPD curve	38 min	15 min
Vinyl ester		
Cellulosic curve	175 min	20 min
Epoxy		
Cellulosic curve	194 min	23 min
Phenolic		
Cellulosic curve	110 min	33 min
NPD curve	72 min	18 min

The fire resistance, as measured by the furnace test, is defined as the time required for the average temperature of the cold face of the panel to reach 160°C (or for a hot spot to reach a temperature of 170°C). The fire resistance times shown in table 11.4 are, as might be expected, significantly shorter than the complete penetration times. One factor to note is that the slow burn-through effect is not particularly resin dependent. Although phenolic resins are shown by the cone calorimeter test to be significantly better in terms of heat release (and are also better in terms of smoke and toxicity), this is not strongly reflected in the indicative test, although the phenolic does give the best result for the NPD tests. In fact all four types of resin are currently being used in thick laminates for heat protection applications offshore and elsewhere.

It is expected that in many areas of offshore use, composite panel components will be required to achieve the stringent H60 or H120 rating, corresponding to resistance values of 60 and 120 minutes in the NPD test. Ways in which these performance values can be achieved will be discussed below.

11.4.1.3 Limitations of furnace tests for composites

Some limitations of the furnace test must be acknowledged. The first is that the test was originally intended for non-combustible materials. Achieving the stipulated rate of temperature increase requires the combustion of a near-stoichiometric mixture of gas and air, with the result that the furnace atmosphere is depleted of oxygen. In addition, compared to a real hydrocarbon fire, the furnace atmosphere is less turbulent and hence the erosive effect is further reduced.

Alternatives to furnace tests are pool fire and jet fire testing. Both have an important place in composite fire testing and are yielding useful results. However, each of these tests has some disadvantages. Pool fire tests can be subject to variability due to atmospheric effects and jet fire tests are very expensive to carry out. In response to the need for further fire testing, a pool fire facility and an improved furnace facility are currently being set up at the University of Salford as part of Phase II of the Marinetech programme [7]. The improved furnace facility is of particular interest, as the furnace will incorporate an additional internal jet which will be used to superimpose some of the features of a jet fire onto the conventional indicative test.

11.4.2 Factors Influencing Fire Performance

It is worth summarising the probable factors contributing to the slow burn-through performance of thick composite laminates in fire. These are:

i) **Transport properties of the laminate.** The fact that the thermal conductivity and diffusivity of FRP are lower than those of steel is clearly an important factor. However, computer simulations of the unsteady state thermal response of panels to fire show that this is not the only effect operating.

ii) **Transport properties of the residual glass.** The

reinforcement, depleted of resin, remaining on the surface of the laminate has a lower thermal conductivity than that of the laminate itself. Computer simulations show that this has a significant effect. Moreover, it can be shown that procedures which help to retain the depleted reinforcement on the panel surface (such as the addition of silica or ceramic fibre plies, which have better refractory properties than E-glass) can further improve fire resistance.

iii) **Endotherm due to decomposition and vaporisation.** The processes of resin decomposition and vaporisation of the decomposition products can be expected to absorb heat within the laminate. Water present in the laminate will also absorb latent heat.

iv) **Convection of volatiles.** As gaseous products pass through the laminate towards the hot surface they can be expected to produce a cooling effect. In addition, when they reach the laminate surface they may form a protective thermal boundary layer. Although these volatiles are flammable, their contribution in a hydrocarbon fire may be insignificant, compared to the thermal release already taking place in the fire.

11.4.3 Sandwich Construction

Twin-skinned construction is particularly convenient and efficient for composite panels. Many partitions and claddings are required to resist both blast and fire loadings. The blast requirement is generally in the form of an assumed uniformly distributed pressure to simulate the blast event. A frequently encountered overpressure design requirement is 0.5 bar, although it has recently been argued that higher pressures, possibly up to 1 bar should be resisted. Arguments for both 0.5 bar and 1 bar specifications have been advanced. It is generally agreed that, although some blasts may generate even larger transient pressures, there is little to be gained from attempting to design panels to resist loadings in excess of 1 bar.

Figure 11.12 shows a comparison between the form of construction currently used for steel blast walls and two types of composite sandwich configuration. In the steel case, the load-bearing member is a single corrugated panel, protected on one side by ceramic wool insulation. With composite sandwich construction the fire performance of thick laminates can be enhanced by the use of fire-resistant core materials. Special consideration is required when choosing the core material, as most conventional load-bearing cores for composite sandwich panels, polymer foams for example, have not been developed with hydrocarbon fire performance in mind. The materials with the best thermal performance, ceramics, tend to be either brittle or to be in a form which is non-structural. The two possible options for panel cores, shown in figure 11.12, are:

i) lightweight, non-structural, ceramic insulating material
ii) structural core material

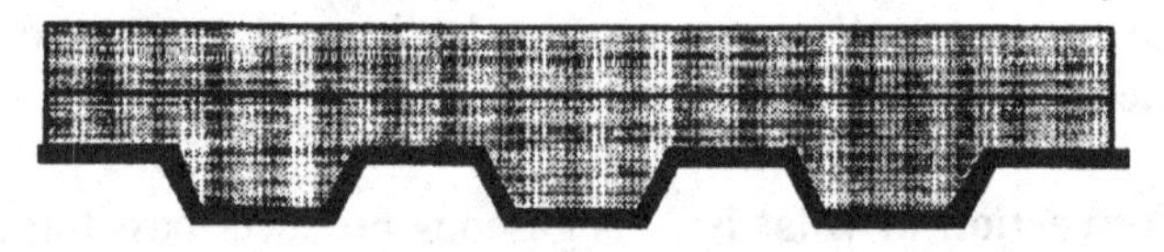

Figure 11.12. Standard form of construction for steel blast panels, compared with the alternative forms of construction for twin skin composite panels.

Examples of option (i) include ceramic blanket insulation, as used presently in steel fire protection panels or even lightweight firebricks. Because this type of core is non-structural, some additional stiffening elements must be incorporated into the panel to take the shear and normal loads normally taken by the core. Examples of the type of element which can be used include pultruded FRP or even metallic sections.

For option (ii), one possibility is to use a conventional structural core material with moderate fire resistance. The best example would be end-grain balsa, which is widely used as a panel core material and which, by virtue of its low thermal conductivity, has modest fire barrier qualities. End-grain balsa is often used in marine FRP bulkheads which require some fire containment capability. This option would give excellent mechanical performance, combined with moderate integrity in a hydrocarbon fire.

The alternative is to use a ceramic or refractory core with load-bearing capability. There are several compressed ceramic or cementitious board materials available, such as for instance, the 'Vermiculux' material supplied by Cape Products. In addition to structural capability, materials of this type have excellent fire penetration properties. The disadvantage is that many of the more attractive candidate board materials have a density penalty. Recently, a number of proprietary phenolic-bonded core materials have been developed, which are rather lighter in weight than the ceramic board option.

11.4.4 Fire-Resistant Properties of Twin-Skin Panels

The Marinetech North West Phase I programme carried out some basic work to quantify the fire performance of composite sandwich laminates. Figure 11.13 shows an experimental panel containing different types of core material, corresponding to cases i) and ii) above. The core materials used were ceramic wool blanket and end grain balsa. In case i) the required structural integrity was achieved by the use of pultruded sections bonded into the panel.

One additional variable was examined: the use of two plies of silica fabric in the laminate, to improve the integrity of the reinforcement after the resin had been depleted. The panel was tested in the indicative fire test, using the hydrocarbon curve.

Figure 11.14 shows the measured cold face temperatures of the panel in the locations corresponding to the different material combinations. It can be seen that all the material combinations examined in this particular test lasted for 60 minutes, those with the ceramic core material lasting in excess of 120 minutes. The presence of the silica fabric does appear to be beneficial.

Figure 11.15 shows an ESDV protection box which uses twin-skinned composite panels to achieve the required blast and fire protection levels. In this case the homogeneous core option was chosen by the manufacturers and pultruded skins were employed.

11.4.5 Design of Twin-Skin Blast and Fire Protection Panels

Most offshore platforms carry a substantial tonnage of steel blast

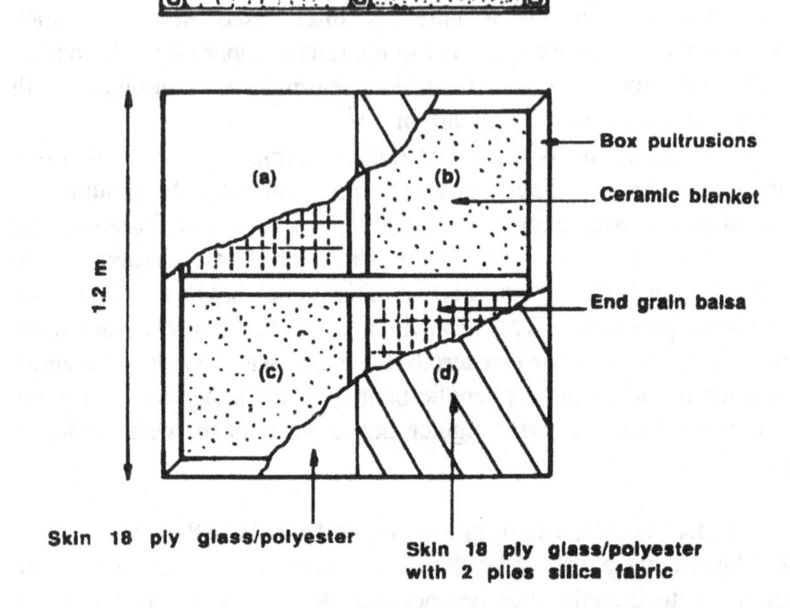

Figure 11.13. Lay-out of test blast/fire protection panel for the evaluation of different types of sandwich and core element in the indicative fire test.
Key: (a) glass/polyester skins + end grain balsa core
(b) glass/polyester skins with added silica fabric + ceramic blanket core
(c) glass/polyester skins + ceramic blanket core
(d) glass/polyester skins with silica fibre + end grain balsa core.

walls and cladding panels with an H60 or H120 requirement. Redesign of these elements in twin-skinned GRP laminates of the type described above could achieve a significant weight-saving.

Consider the design of a panel, in polyester/woven glass to the following requirement:

Nature of Support: simply supported with a span of 2.5 metres.
Blast Loading: a defined level of overpressure, assumed to be acting as a uniformly distributed load.
Fire Loading: panel must withstand H120 furnace test.
Materials Data: a laminate of polyester/glass woven rovings has the following properties:

Tensile strength	280 MPa
Young's modulus	18 GPa

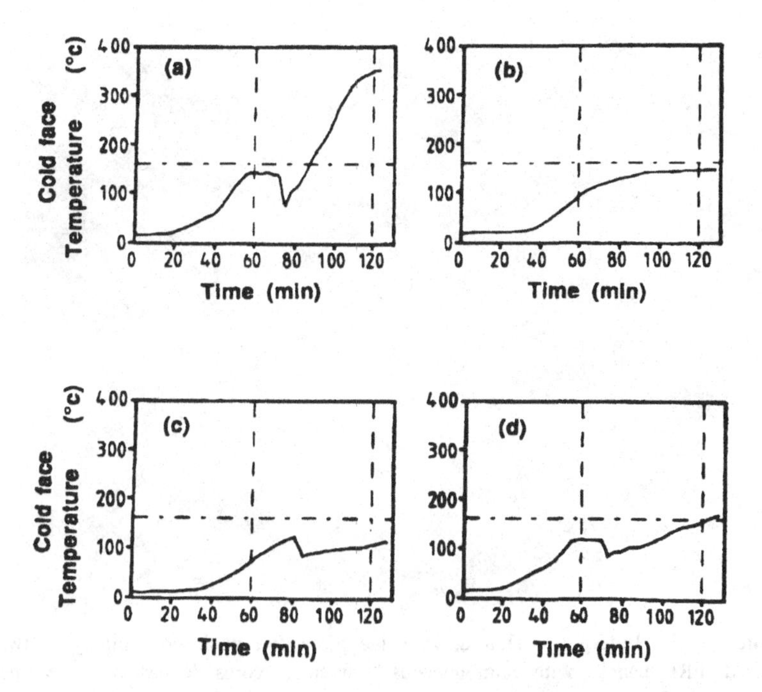

Figure 11.14. Rear face temperature of blast/fire protection panel in the indicative fire test (NPD curve) measured at four different locations. Key as in figure 11.13.

The option of using a non-structural core material, with bonded-in pultruded reinforcing elements will be chosen for this illustration, as this generally leads to a lighter design than the homogeneous core option. To pass the H120 fire test with a non-structural ceramic fibre wool requires a minimum skin thickness of 9 mm, with a core of at least 35 mm thickness.

Consider first the mechanical requirement. The maximum bending moment per unit width due to blast load will be

$$M_{max} = PL^2/8 \tag{11.1}$$

where P is the overpressure and L the span.

Considering only the load borne by the skins of the panel, the bending moment per unit width, M, can be related to the skin stress, σ, by

$$M = \sigma . t . h \tag{11.2}$$

where t and h are the skin thickness and the distance apart of the skin

Figure 11.15. Emergency shut down valve protection enclosure employing twin skinned FRP panels with homogeneous refractory cores (Courtesy of Vosper Thornycroft Ltd).

centres. This simple formula, which is accurate enough for present purposes, assumes that the skin thickness is small, compared with h. Combining these equations gives

$$P = 8\sigma_{max}.t.h/L^2. \tag{11.3}$$

Using an allowable stress of 280 MPa in equation (11.3), along with the dimensions needed to give the required fire performance (t = 9 mm and h = 44 mm) gives the overpressure which an H120 panel would fail as 1.42 bar. This is well in excess of the usual design requirement for panels of this type.

With the non-structural core design it is always necessary to avoid the possibility of premature compressive skin failure due to periodic buckling between the pultruded reinforcing elements. Roark [22] gives the following expression for the buckling of a plate with edge restraints:

$$\sigma_{max} = KE(t/b)^2/(1-\nu^2) \tag{11.4}$$

where E is Young's modulus, ν, Poisson's ratio and b, the distance between restraints. For the type of restraint offered by the bonded internal members, with $L/b > 3$, Roark [22] gives $K = 5.73$.

The ratio of support spacing to skin thickness needed to allow the material strength of 280 MPa to be approached can be found from equation (11.4). Assuming a Poisson's ratio value of 0.3 and solving for b/t gives a ratio of 20.1 for this quantity. (Factors related to the buckling of reinforced FRP panels are discussed at greater length in the book by Smith [23]). In fact, in the present design, some weight could be saved by using a larger ratio than this, since the panel performance for a maximum stress of 280 MPa is well in excess of the blast requirement.

For the case of a skin thickness of 9 mm, with a 35 mm thick kaowool core and internal elements consisting of box section pultrusions at 200 mm spacing the total panel weight would be 42 kg/m^2. An equivalent steel panel for similar loading would, by contrast, weigh at least 55 kg/m^2.

11.5 CONCLUSIONS

Significant factors in favour of the use of composites offshore are demonstrable improvements in reliability in the case of pipes, and the possibility of a significant role in fire protection in the case of panels. Work on overcoming the negative factors, lack of design codes, fabrication limitations and lack of quantified data on fire performance continues.

The laminate thickness effect in fire can be used synergistically in twin-skin panel designs using refractory core materials. In such designs the composite outer skins provide the initial mechanical and blast performance along with a useful measure of initial fire penetration resistance. The refractory core then provides substantial further fire resistance to ensure the required degree of fire protection. This form of construction offers weight-savings compared with steel blast panels.

11.6 ACKNOWLEDGEMENTS

The author, who is involved with the Marinetech North West programme on Cost-effective Use of Fibre Reinforced Composites Offshore would like to thank the sponsors of that programme for their support and for permission given to publish some of the results presented here: Admiralty Research Establishment, AGIP (UK), Amoco Research, Balmoral Group, BP Exploration, BP Research, British Gas, Ciba-Geigy, Conoco, Dow Rheinmunster, Elf (Aquitaine),

Elf (UK), Enichem (SPA), Exxon, Fibreforce Composites, Kerr McGee Oil (UK) Ltd., MaTSU, Marine Technology Directorate Ltd., Mobil Research and Development, Mobil North Sea Ltd., UK Ministry of Defence (Navy), UK Offshore Supplies Office, Phillips Petroleum, Shell Expro, Statoil, Total Oil Marine, UK Department of Energy, V.S.E.L. and Vosper Thornycroft. The research programme involves the Universities of Glasgow, Liverpool, Salford, UMIST and Newcastle upon Tyne.

11.7 REFERENCES

1] Study on Lightweight Materials for Offshore Structures. May 1987. *Project Report Fulmer Research/Wimpey Offshore.* Report no. WOL 161/87.

2] A study for the Development of the Use of Aluminium and FRP Composite Materials for Living Quarters. *Project Report Offshore Design Engineering,* sponsored by OSO, Phillips Petroleum, Total and Conoco.

3] Stokke, R., Glass Fibre Reinforced Plastics (GRP) Offshore, *Reports of Multi-Sponsor Programme.* Centre for Industrial Research, Forskningsveen 1, Postboks 350, Blindern, 0314 Oslo 3, Norway.

4] *Programme to study FRP in Accommodation Modules and Ship Superstructures.* ODE/BMT.

5] Godfrey, P.R., Davis, A.G., 'The Use of GRP Materials in Platform Topsides Construction and the Regulatory Implications', Proc. 9th Intl. Conf. *Offshore Mechanics and Arctic Engineering,* **3**(A), Houston, Texas, February 1990. pp 15-20.

6] Gibson, A.G., Spagni, D.A., Turner, M.J., "The Cost-Effective Use of Fibre Reinforced Composites Offshore", Multi-Sponsor Research Programme, Phase I Final Report, Marinetech North West, Coupland III Building, The University, Manchester M13 9PL.

7] Gibson, A.G., Spagni, D.A., "The Cost-Effective Use of Fibre Reinforced Composites Offshore", Phase II Prospectus, Marinetech North West, Coupland III Building, The University, Manchester, M13 9PL.

8] Williams, J.G., "Developments in Composite Structures for the Offshore Oil Industry", paper OTC 6579, Proc. 23rd Annual Offshore Technology Conf., Houston, Texas, USA, May 1991.

9] Ashby, M.F., Jones, D.R.H., "Engineering Materials, and

Introduction to their Properties and Applications", Pergamon Press, Oxford, 1980.

10] Waterman, N.A., Ashby, M.F., (ed.), "Elsevier Materials Selector", Elsevier Applied Science, Barking, Essex, 1991.

11] Grim, G.C., "The Use of GRP Piping in the Oil and Ballast Water Systems of the Draugen Gravity Base Structure", Proc. Intl. Conf. *Polymers in a Marine Environment*, IMarE, London, October 1991.

12] Forsdyke, K.L., "Phenolic Matrix Resins - The Way to Safer Reinforced Plastics", paper 18-C, Proc. 43rd Annual Conf. Society of The Plastics Industry, Cincinnati, Ohio, February 1988.

13] Quinn, J.A., "Pultrusion: An Economic Manufacturing Technique", Metals & Materials, May 1989. pp. 270-273.

14] Quinn, J.A., "Design manual for Engineered Composite Profiles", Fibreforce Composites Ltd., Fairoak Lane, Whitehouse, Runcorn, Cheshire, WA7 3DV, 1988.

15] Engelen, H., 'The Influence of the Process Parameters on the Mechanical Properties of Pultruded GRP Profiles', 24th Journees Europeenes des Composites, JEC, Centre de Promotion des Composites, Paris, 1989,

16] Grim, G. C., "Shipboard Experience with Glass-Reinforced Plastic (GRP) Pipes in Shell Fleet Vessels", Proc. Intl. Conf. *Polymers in a Marine Environment*, IMarE, London, October 1987.

17] Guiton, J., "An All-GRP Piping, Support and Walkway System for Tanker Weather Deck Applications", Proc. Intl. Conf. *Polymers in a Marine Environment*, IMarE, London, October 1987.

18] Marks, P.J., "The Fire Endurance of Glass-Reinforced Epoxy Pipes", Proc. Intl. Conf. *Polymers in a Marine Environment*, IMarE, London, October 1987.

19] Grim, G.C., Twilt, L., "Fire Endurance of Glass Fibre Reinforced Plastic Pipes Onboard Ships", Proc. Intl. Conf. *Polymers in a Marine Environment*, IMarE, London, October 1991.

20] Ciraldi, S., Alkire, J.D., Huntoon, G., "Fibreglass Firewater Systems for Offshore Platforms", paper OTC 6926, Proc. 23rd Annual Offshore Technology Conf., Houston, Texas, May 1991. pp 477-484.

21] Hume, J., "Assessing the Fire Performance Characteristics of

Composites", Proc. Conf. *Materials and Design against Fire*, IMechE, London, October 1992.

22] Young, W.C., (ed): "Roark's Formulas for Stress and Strain", 6th edition, McGraw-Hill, New York, 1989.

23] Smith, C.S., "Design of Marine Structures in Composite Materials", Elsevier Applied Science, Barking, Essex, UK, 1990.

12 REGULATORY ASPECTS IN DESIGN

12.1 APPLICABLE RULES

12.1.1 Background: IMO Code for Dynamically Supported Craft

All passenger ships in international trade carrying more than 12 passengers must meet the requirements of the International Convention for Safety of Life at Sea (SOLAS). The SOLAS requirements are issued by the International Maritime Organisation (IMO) and represent international agreement between the contracting governments. In addition, the IMO has introduced special requirements for cargo ships and tankers, and many other recommendations which governments are invited to use as national requirements.

The IMO has also seen the need for regulations taking care of ships which embody features of a novel kind. This led to the adoption by the IMO of Resolution A.373(x), the "Code of Safety for Dynamically Supported Craft", which was published in 1978.

The provisions of the Code are based on the following concepts:

- that the distances covered and the worst intended environmental conditions in which operations are permitted will be restricted
- that the craft will at all times be in reasonable proximity to a place of refuge
- that specified facilities will be available at the base port from which the craft operates
- that the Administration is able to exercise strict control over the operation of the craft
- that rescue facilities can be rapidly provided at all points in the intended service
- that all passengers are provided with a seat and no sleeping berths are provided
- that facilities are provided for rapid evacuation into suitable survival craft

At present the Code applies basically to craft which

- carry more than 12 but not over 450 passengers, with all passengers seated
- do not proceed in the course of their voyage more than 100 nautical miles from the place of refuge

Craft having special category spaces intended to carry motor vehicles with fuel in their tanks, but subject to the above limitations, are also covered. Additionally, it is accepted that all or parts of the Code may also be applicable to craft exceeding these limits.

The Code allows the hull and superstructure to be designed for loadings appropriate to the operating conditions to which the craft will be restricted. While stating that the hull should be constructed of approved non-combustible materials having adequate structural properties, it nevertheless allows the use of other materials provided additional precautions taken are sufficient to ensure an equivalent level of fire safety.

The Code is currently undergoing a major revision. The aim is to establish a safety standard that will be fully accepted as an equivalent alternative to the SOLAS convention, and to extend the application to vessels with larger operating ranges and numbers of passengers.

The IMO Code has a major bearing on the requirements of the Classification Societies with regard to high speed and light craft.

12.1.2 Rules of the Classification Societies

The main rules for classification of ships that are issued by the major Classification Societies, such as Lloyd's Register, American Bureau of Shipping, Bureau Veritas and Det Norske Veritas (DNV), apply principally to steel ships. As an alternative, DNV introduced in 1972 the Tentative Rules for the Construction and Classification of Light Craft. The 1985 DNV Rules for Classification of High Speed Light Craft were a development of the 1972 Tentative Rules and were formulated in the spirit of IMO A.373(x).

The current Tentative Rules for Classification of High Speed and Light Craft [1] were issued in 1991 and are a further development of the 1985 Rules. The Rules are constantly under review since this is a field where the range of application (in terms of vessel types, sizes and speeds) is rapidly expanding.

DNV has also previously published Rules for Construction and Certification of Vessels Less than 15 metres. These Rules applied to craft which do not require classification as such, and were an English translation of Nordic Boat Standard 83. In 1990 the Nordic Boat Standard was revised, and this now forms the basis for DNV certification of vessels less than 15 metres. However, the new Standard has not so far been translated into English and its future is uncertain because of likely developments in EC standardisation.

In the following discussion, attention will be focused on the DNV Rules, and in particular on the 1991 Tentative Rules for High Speed and Light Craft.

12.1.3 Main Class

The main rules for classification of steel ships specify technical requirements for hull structures, machinery installations and equipment with respect to strength and performance. The requirements are intended to safeguard against hazard to the vessel, personnel and environment by reducing the risk of structural damage, machinery damage, fire and explosions and other incidents which may be caused by technical failures.

For steel ships the main class **+1A1** is an approval, survey and supervision system for vessels being built worldwide. This system establishes the standard of main constructional and system elements necessary for a vessel to be safe at sea. The main class ensures that a vessel will be built of suitable materials, by adequate workmanship and with necessary strength, stability and fire safety. Vital installations such as engines, fuel systems, bilge system, electrical installations etc. are required to correspond with specifications as stipulated in the main class. In this way the main class ensures that the basic vessel is delivered in a condition such that it can operate satisfactorily for many years. During each main class period, normally 4-5 years, the vessel is checked regularly, the maintenance is inspected, and possible damage repairs are supervised.

Governmental bodies worldwide require the main class certificate to be valid as the basis for the issuance of a national or international trade certificate and register.

The notation **Light Craft (LC)** was first introduced as an additional class notation by the DNV Tentative Rules for the Construction and Classification of Light Craft of 1972 and continued in the 1985 Rules. It applies to vessels for which

$$\Delta \leq (0.13\ L\ B)^{1.5} \tag{12.1}$$

where Δ is the displacement in tonnes, L is the length between perpendiculars in metres and B is the overall breadth in metres (excluding the tunnel breadth in the case of a catamaran).

The 1991 Tentative Rules introduced the additional definition of a high-speed vessel as one for which

$$Fn > 0.7 \tag{12.2}$$

where Fn is the Froude number $v/\sqrt{(g_0 L)}$, v is the maximum service speed in m/s, and $g_0 = 9.81\,m/s^2$. If V is the maximum service speed in knots, this is equivalent to

$$V/\sqrt{L} > 4.2 \tag{12.3}$$

Vessels satisfying the conditions on both Δ and Fn have the notation **High Speed Light Craft** (**HSLC**) while those satisfying only the condition on Δ have only the notation **LC**. In the 1991 Tentative Rules, these are now part of the main class.

Ordinary classification rules are based on service experience from the steel ship merchant fleet. The associated loadings are too conservative for lightweight craft. Instead of being exposed to full sea loading, light craft to some extent respond away from the loading. The slamming loads occur over a very short period of time and only part of the load will create stresses in the construction. Because of this it is possible to reduce the scantlings compared to ordinary ships. Then, with lightweight materials such as aluminium and fibre reinforced plastics, and with light cargo or only passengers on board, higher speeds can be attained.

With the **+1A1 LC** and **+1A1 HSLC** class notations the craft is designed for specified acceleration levels and restricted sea state/speed conditions to avoid structural overloading and unacceptable levels of passenger discomfort. The notations are only applied in connection with a specified Service Restriction.

12.1.4 Service Restriction R0-R4

For steel ships the **+1A1** main class is assigned to vessels operating worldwide. These vessels shall be capable of withstanding all sea state conditions encountered on the oceans, and in addition have spare parts on board for minor repairs and normal maintenance.

Vessels operating on a certain route or within a restricted distance from a base port do not need to carry on board such a complete complement of maintenance equipment and spare parts. Most maintenance and repair work will be carried out in the base port.

When a vessel is operating within a short distance from a safe harbour, it is possible to decide on a restricted range of sea states in which the vessel shall operate. Thus reduced strength requirements may be justified. Furthermore, vessels operating in sheltered waters, or within restricted sea or wind conditions, may operate with less anchor and mooring equipment.

These vessels are assigned an additional service restriction notation currently between **R0** and **R4**, which determines the maximum distance the vessel will be permitted to operate from the nearest harbour or safe anchorage, as shown in table 12.1. For high speed and light craft with the additional class notation **Passenger** (see section 12.1.5), the restrictions **R1** to **R4** are applicable, but distances must not exceed those shown against the restriction **RX** in table 12.1.

Table 12.1. Service restrictions. Maximum distances from nearest harbour or safe anchorage, in nautical miles.

Service notation	*Season*		
	Winter	*Summer*	*Tropical*
R0	300	No restriction	No restriction
R1	100	300	300
R2	50	100	250
R3	20	50	100
R4	Enclosed fjords, rivers, lakes	20	50
RX	VL/20	VL/10	VL/7

12.1.5 Other Class Notations

Vessels arranged and/or strengthened for a special service and meeting specified requirements may be assigned one of the following special service or type notations:

- Passenger
- Cargo
- Car ferry
- Patrol
- Yacht

Other voluntary additional class notations available for high speed and light craft are:

- Periodically Unattended Machinery Space (E0)
- Nautical Safety Class (NAUT)
- Dynamic Positioning System (DYNPOS)

Earlier additional class notations Stability and Floatability (SF) and Fire Safety Class (F) are now replaced by requirements in the main class and special service or type notations.

12.2 DESIGN LOADS FOR HIGH SPEED AND LIGHT CRAFT

12.2.1 Introduction

Design loads for high speed and light craft are conveniently divided into global and local loads. Broadly speaking, for a monohull vessel the local loads are used in the design of smaller structural elements such as individual stiffened panels and frames, while the global loads are used for assessing the overall strength of the hull girder. For a vessel with twin hulls the bridging structure between the hulls is subjected to a transverse bending moment and shear force arising from the resultant forces and moments on the hulls. These loadings are of a global nature similar to hull girder moment and shear, and in a simple treatment the design of a bridging structure can be based on these two quantities. (However, in practice the transverse frames of a twin-hull structure are often designed on the basis of a frame analysis with a series of load cases that also includes likely combinations of local loadings.) A further global loading for twin-hull vessels is the so-called pitch-connecting moment, which represents a torsional loading on the bridging structure.

The loads to be assumed in the design, both local and global, are to a large extent stipulated in the rules of the Classification Societies. However, the loads from the rules often need to be either supplemented with, or in some cases replaced by, more detailed data from hydrodynamic analyses or model tests. In some cases the loads calculated from rule formulae are found to be appreciably conservative.

In this section, the state of affairs as far as standard rule requirements for light, planing or semi-planing craft, will be presented, referring particularly to the 1991 DNV Tentative Rules for High Speed and Light Craft.

Hitherto the Bureau Veritas [2] and American Bureau of Shipping [3,4] Rules have had a number of similar features to the DNV Rules, but have been somewhat simpler in many areas. Provisional Rules recently issued by Lloyd's Register of Shipping for high speed catamarans [5], while applying only to aluminium alloy vessels, also have common features in relation to design loads.

12.2.2 Vertical Acceleration

The design vertical acceleration at the LCG, a_{cg}, is currently given as a function of vessel application and $V/\sqrt{L}$ ratio (which represents the Froude number Fn). This is under review and the design acceleration will probably also become a function of the service restriction **R0** - **R4**. For passenger vessels, the design a_{cg} is generally $1.0\,g_0$ where g_0 is the gravitational acceleration.

The design vertical acceleration a_v at any other position than the LCG is given by

$$a_v = k_v a_{cg} \tag{12.4}$$

where k_v is given by figure 12.1.

The design vertical acceleration is generally related to the mean of the 1/100 highest accelerations encountered.

The vertical acceleration at the LCG should not exceed the specified design value. If a_{cg} is found to be greater than the specified design value for the craft's maximum speed at maximum significant wave height, this implies a speed restriction for rough conditions. Thus a limiting speed/wave height relationship must be specified for a given craft. The 1991 Tentative Rules give a formula based on that derived by Savitsky and Brown [6]. This takes account of the trim and deadrise angles. A second formula relating to vessels operating in displacement mode is applied for low speeds.

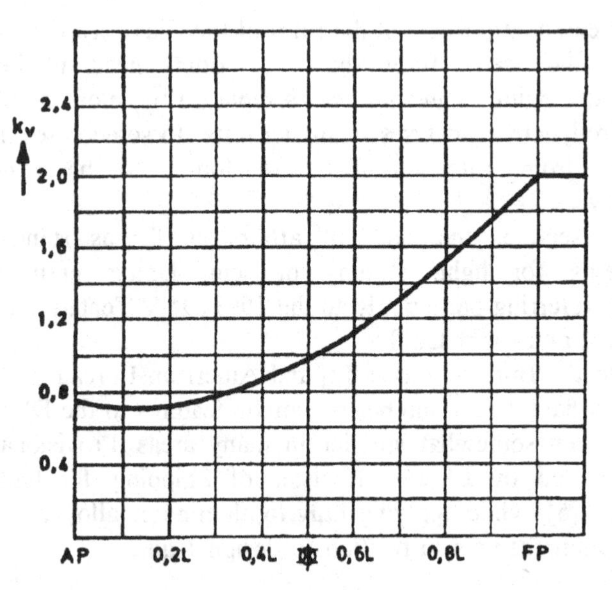

Figure 12.1. Longitudinal distribution factor for design vertical acceleration, k_v (under review).

The requirement for speed reduction is given in an appendix to the classification certificate and must be displayed on board the craft. An alternative approach is to install an accelerometer at the LCG and require that the craft shall not be operated so as to exceed a specified value of measured a_{cg}.

12.2.3 Local Loads

12.2.3.1 Slamming pressure on bottom

The DNV Rules specify a design pressure for slamming as follows:

$$p_{sl} = 1.3\,k_l(\Delta/A)^{0.3}\,T^{0.7}\,[(50-\beta_x)/(50-\beta_{cg})]\,a_{cg}\ \text{kN/m}^2 \qquad (12.5)$$

where

- a_{cg} = design vertical acceleration as defined previously
- k_l = longitudinal distribution factor for pressure from figure 12.2
- Δ = fully loaded displacement in tonnes salt water
- A = design load area for element considered, in m^2
- T = draught at L/2 in m
- β_x = deadrise angle in degrees at middle of load area considered (minimum 10°, maximum 30°)

β_{cg} = deadrise angle in degrees at LCG (minimum 10°, maximum 30°)

Note that the variation of the factor k_l is currently under review as the form shown in figure 12.2 is believed to lead to underestimation of the slamming pressure in the forward part of the craft in some cases.

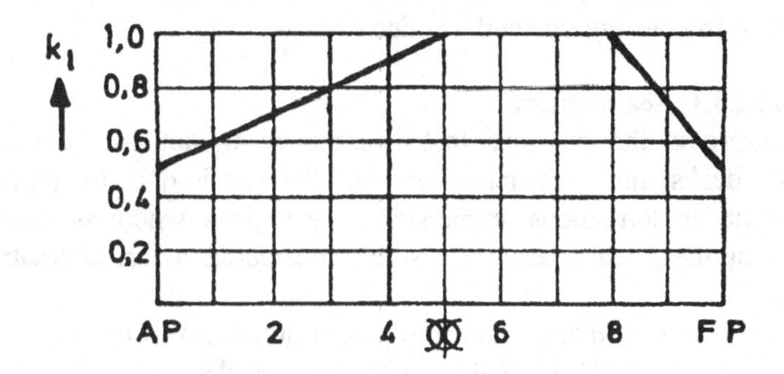

Figure 12.2. Longitudinal pressure distribution factor for slamming, k_l.

The above requirement applies to the region from the keel line up to the chine, upper turn of bilge or pronounced sprayrail and is based closely on the approach by Allen and Jones [7]. The slamming pressure is an equivalent static pressure assumed to apply over the design area A. The dependence on design area A reflects the fact that slamming loads are in reality concentrated over limited areas, so that when a larger area of structural element is being considered the equivalent distributed pressure due to slamming will be smaller.

For plating, and for laminates in single skin FRP construction, the design area A is usually to be taken as $2.5\,s^2$, where s is the stiffener spacing. Generally A need not be taken as less than $0.002\ \Delta/T$.

An additional formula (currently under review) is given for the slamming pressure in the forward part of the craft due to pitching when operating at low speed in the displacement mode.

12.2.3.2 Forebody side and bow impact pressure; slamming pressure on flat cross-structure

In the current DNV Tentative Rules a formula is provided to give design slamming pressures in the forward part of the vessel above the waterline or the chine or upper turn of bilge (generally taken as the point where the deadrise angle is 70°), whichever is the lower. It applies up to a distance 0.4 L from the forward perpendicular. The pressure is a function of the flare angle and the angle between the water plane/hull intersection line and the longitudinal line in the region considered. It is also explicitly dependent on the longitudinal position.

Formulae are also given for the design slamming pressure on flat cross structures (such as catamaran tunnel tops), and for the height above which slamming does not need to be considered.

These formulae are currently under review.

12.2.3.3 Sea pressure

Sea pressure is the pressure that is assumed to act on the sides, exposed decks and superstructure of the craft due to partial, intermittent or continuous immersion. For regions where slamming pressure applies, sea pressure must be considered as an alternative loading.

The sea pressure, like the slamming pressure, is given by empirical formulae in the DNV Rules. For the craft's side including superstructure side and weather decks these are as follows.

For load point above summer load water line, but above chine, turn of bilge or pronounced spray rail:

$$p = 10\ h_0 + (k_s - 1.5\ h_0/T)\ 0.08\,L\ \text{kN/m}^2. \tag{12.6}$$

For load point above summer load waterline:

$$p = a\ k_s\ (cL - 0.53h_0)\ \text{kN/m}^2 \tag{12.7}$$

subject to the limit

$p_{min} = 6.5$ kN/m² for craft's sides
or $p_{min} = 5$ kN/m² for weather decks

(These lower limits are currently under review.)

In the above,

h_0 = vertical distance in m from the waterline at draught T to the load point
k_s = 7.5 aft of midships
= $5/C_B$ forward of F.P. (where C_B = block coefficient = $\Delta/(1.025\ L\ B\ T)$)
with linear variation between, as figure 12.3

a = 1.0 for craft's sides and open freeboard deck
= 0.8 for weather decks above freeboard deck
c = factor for service restriction notation:
0.080 for restrictions **R0**, **R1** and **R2**
0.072 for restriction **R3**
0.064 for restriction **R4**

Different minimum pressures apply to superstructure end bulkheads and deckhouses.

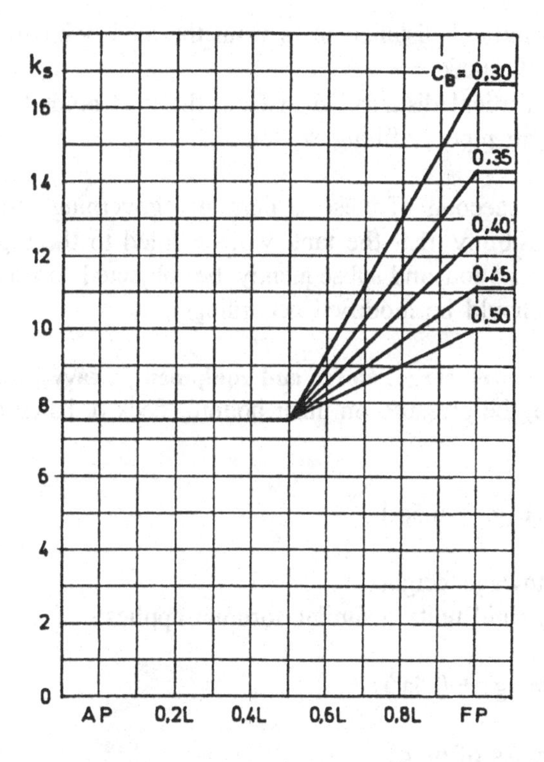

Figure 12.3. Sea load distribution factor for sea pressure, k_s.

Note that, because of the factor c, the sea pressure is smaller for service restrictions **R3** and **R4** than for **R0**, **R1** and **R2**.

In the 1991 Tentative Rules an alternative set of formulae is given for low-speed operation in the displacement mode, but these are currently under review.

Formulae are also given for pressure on watertight bulkheads for condition with flooded compartment, and for pressure on the inner bottom for a case with flooded double bottom.

12.2.3.4 Liquids in tanks

The pressure in tanks is to be taken as the greatest of the following:

$$p = \rho\,(g_0 + 0.5a_v)\,h_s \quad kN/m^2 \qquad (12.8a)$$
$$p = 0.67\,\rho g_0 h_p \quad kN/m^2 \qquad (12.8b)$$
$$p = \rho g_0 h_s + 10 \quad kN/m^2 \text{ for } L \leq 50\,m \qquad (12.8c)$$
$$p = \rho g_0 h_s + 0.3L - 5 \quad kN/m^2 \text{ for } L > 50\,m \qquad (12.8d)$$

in which

h_s = vertical distance in m from the load point to the top of the tank

h_p = vertical distance in m from the load point to the top of air pipe or filling station

In practice the second of these is normally governing. Note that, if there is a possibility that the tank will be filled to the top of an air pipe or filling station and subsequently be subjected to accelerations, the pressure should be modified accordingly.

12.2.3.5 Dry cargo, stores and equipment; heavy units

For dry cargo, the pressure on inner bottom, deck or hatch cover is to be taken as

$$p = \rho H\,(g_0 + 0.5a_v) \qquad (12.9a)$$

where H = stowage height.

For heavy, rigid units a similar formula applies:

$$P_v = M\,(g_0 + 0.5a_v) \qquad (12.9b)$$

where M = mass of unit.

Note that both of these expressions imply a static component plus a dynamic component which is based on an effective vertical acceleration $0.5a_v$.

12.2.4 Global Loads (Hull Girder Loads)

12.2.4.1 Hull girder bending and shear

The DNV Rules indicate that, for craft with "ordinary hull form" having L/D < 12 (where L is length between perpendiculars and D is moulded depth) and L < 50 m the minimum strength standard for the hull girder is normally satisfied as a result of meeting the local strength requirements, i.e. it is local strength requirements that govern.

However, for other types of craft, craft with L/D > 12 and craft with L > 50 m the longitudinal strength must be checked for the loadings defined in figure 12.4. The crest landing condition is one in which the hull girder is out of the water, being accelerated upwards as a consequence of a slamming load centred at the LCG. The weight distribution of the hull girder is thus increased by the acceleration at the LCG. The slamming pressure is considered to be applied to an area equal to the reference area A_R centred at the LCG, where

$$A_R = 0.6\ \Delta\ (1 + 0.2\ a_{cg}/g_0)\ /\ T \qquad m^2. \tag{12.10}$$

The longitudinal extent l_s of the slamming reference area A_R is given by

$$l_s = A_R\ /\ b_s \tag{12.11}$$

where b_s is the hull breadth at the position concerned (excluding the tunnel in the case of a catamaran).

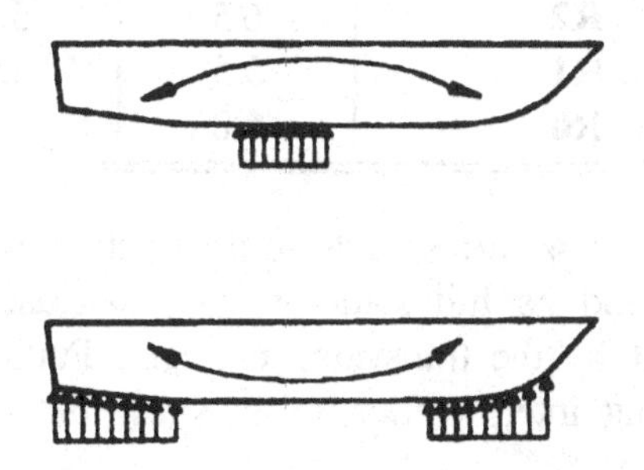

Figure 12.4. Global loadings on hull girder, for longitudinal strength.

The hollow landing case is analysed similarly, but the reference area is divided equally between regions adjacent to the FP and AP.

Simplified formulae are given for estimating the maximum bending moments for the two cases (hogging and sagging, respectively) if detailed information about the mass distribution is not available.

Special consideration of bending moments in a slowed-down condition may be required for special types of vessel such as hydrofoils, ACV's, SES and other small water plane area side wall craft. This may also be required for craft having a low Froude number at full operating speed.

12.2.4.2 Transverse bending moment and shear force for twin-hull vessels

The DNV Rules specify the following values of transverse bending moment M_s and shear force S (figure 12.5) to be assumed for the bridging structure in the absence of more detailed information:

$$M_s = \Delta\, a_{cg}\, b / s \tag{12.12}$$

$$S = \Delta\, a_{cg} / q \tag{12.13}$$

where b = transverse distance between centrelines of hulls and s, q are factors given in table 12.2

Table 12.2. Factors s, q for calculating transverse bending moment and shear force.

Service restriction	s	q
R4	8.0	6.0
R3	7.5	5.5
R2	6.5	5.0
R1	5.5	4.0
R0	4.0	3.0

The limiting case s = 4 corresponds to a condition where one hull is out of the water and its full static weight, increased by the factor a_{cg}/g_0, is supported by the transverse moment. Putting q = 2 would give a corresponding interpretation of the shear, but in fact the lowest value specified is 3.

Additional formulae are provided to cover the horizontal split force and corresponding transverse moment M_s for cases with operation at low speed in the displacement mode.

In practice it is common to perform additional analysis of transverse frames with combinations of local loads.

12.2.4.3 Pitch connecting moment for twin-hull vessels

The pitch connecting moment may be assumed to be

$$M_p = \Delta\, a_{cg}\, L / 8 \tag{12.14}$$

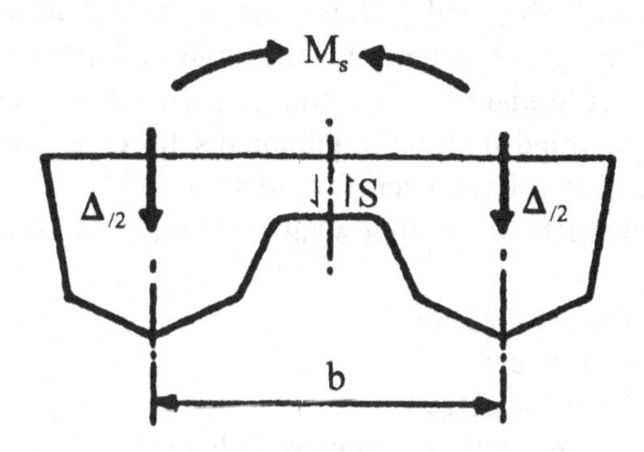

Figure 12.5. Global transverse loading on bridging structure.

When $a_{cg} = g_0$ this corresponds to a docking condition in which the hulls are supported at two points located a distance L/4 fore and aft of the LCG, as shown in figure 12.6.

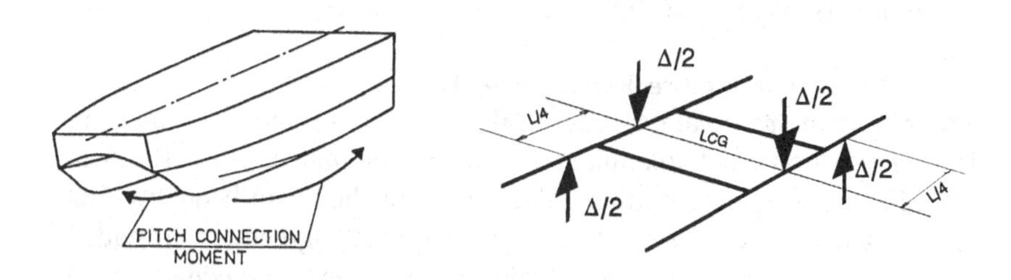

Figure 12.6. Pitch connecting moment on bridging structure.

The introduction of an additional case consisting of a torsional moment about the longitudinal axis of a twin-hull vessel is being considered.

12.3 APPROVAL OF MATERIALS FOR GRP CONSTRUCTION

12.3.1 Introduction

The primary materials of interest for fast craft have hitherto been aluminium alloys and glass-reinforced plastic (GRP). The 1991 DNV Tentative Rules for Classification of High Speed and Light Craft also cover craft constructed in steel. In the case of GRP, construction methods involving both single skin and sandwich are catered for. In the latter case a further material, the sandwich core material, is present. For small high speed craft the core materials are often of end-grain balsa and honeycomb type; for moderate to large size passenger craft rigid foam core materials are almost exclusively used. It is with these materials in mind that the requirements for core materials in the DNV Tentative Rules have been formulated.

For GRP construction the following products require approval:

- Glass fibre reinforcements
- Polyester products
- Sandwich core materials
- Sandwich adhesives and cement (filler)

It is also accepted that alternative materials to glass and polyester may be used, subject to testing and approval in each case.

Approval of the products listed above is generally given as a type approval, valid for a period of four years. The approval is subject to satisfactory results of, among other things, random inspection. Certain requirements have to be fulfilled for each delivery.

12.3.2 Glass Fibre Reinforcements

The requirements relate to chemical composition, moisture content, loss on ignition, and tolerance on weight per unit area (± 10 % of manufacturer's nominal value). They relate explicitly to E-quality, but other qualities are permitted subject to special agreement provided their mechanical properties and hydrolytic resistance are equally good or better.

12.3.3 Polyester Products

The approval of polyester is divided into two different quality grades:

Grade 1: Quality with good water resistance
Grade 2: Quality with normal water resistance

Requirements for resin in liquid condition cover the following:

- density
- viscosity
- acid value
- monomer content
- mineral content
- gel time

- linear curing shrinkage

With the exception of mineral content, these are all expressed as permitted deviations from the manufacturer's nominal values.

For cured resin the following are specified, with different requirements for the two grades:

- density (required to be manufacturer's nominal value)
- hardness measured according to ASTM D 2583-67
- heat deflection temperature (HDT) measured to ISO 75-1974
- water absorption measured to ISO/R 62-1980
- tensile strength measured to ISO/R 527-1966
- modulus of elasticity measured to ISO/R-1966
- fracture elongation

Additional delamination tests are required for polyesters containing waxes or other substances that might lower the secondary bonding ability. These consist of tests to ASTM C 297 on specimens consisting of a primary laminate on top of which a secondary laminate has been built.

12.3.4 Sandwich Core Materials

The DNV Tentative Rules include the following general requirements for core materials:

- They must have stable long-term properties; this may need to be documented.
- On delivery the surface must be such that no further machining is necessary to obtain proper bonding of the material.
- They must normally be compatible with resins based on polyester, vinyl ester and epoxy.

Additional requirements may be introduced for core materials of special composition.

The approval of sandwich core materials is separated into two different quality grades:

Grade 1: Required quality for hull construction
Grade 2: Required quality for less critical applications

Requirements for core materials in the two respective grades cover the following:

- tensile strength and modulus
- compressive strength and modulus (at both 23°C and 45°C)
- shear strength, elongation and modulus
- water absorption
- water resistance
- density
- oxygen index

The values of density and compressive strength (23°C) must be demonstrated for each delivery. For all the listed properties apart from the compressive properties at 45°C and the water absorption and resistance, minimum values are to be given by the manufacturer and verified by approval testing, and specified on the type approval certificate.

Minimum core shear and compression strength requirements for the main hull structural members are given in addition in the part of the Rules that deals with structural design.

12.3.5 Sandwich Adhesives and Cement

Requirements for sandwich adhesives and cement relate to viscosity and linear curing shrinkage for the uncured material, and to tensile strength and fracture elongation, shear strength and water resistance for the cured material.

For a specific core and adhesive combination to be used in a classed vessel it may be necessary to verify that the core adhesive has no adverse effect on the performance of the complete sandwich panel. Static or fatigue tests with four-point bending are normally employed for such verification.

12.4 DESIGN/ANALYSIS OF STRUCTURE - GRP

The DNV Tentative Rules for Classification of High Speed and Light Craft include requirements for both single skin and sandwich construction.

12.4.1 Material Properties and Testing

As mentioned in section 12.3, the DNV Rules define two grades of polyester products, according to their water-resistance properties. Of these, only Grade 1 is allowed to be used in the hull shell laminate for single skin construction, and in the hull outer skin laminate in sandwich construction.

Core materials are also divided into two grades. Only Grade 1 can be used in the hull panels.

Polyester products and core materials are type approved in relation to one or other of these grades. Reinforcements and core adhesives are also type approved.

Strength calculations are to be based on mechanical properties obtained from testing of representative sandwich panels and laminates with respect to production procedure, workshop conditions, raw materials, lay-up sequence, etc. For yards that already have experience of the particular construction method to be used, data from previous tests may be available for use in design calculations. Otherwise special tests need to be carried out for this purpose.

Additionally, testing must be carried out to confirm the adequacy of the actual construction. This testing is usually performed on pieces actually removed from the hull structure, often from cut-outs for sea-water intakes or other piping or ducting.

A minimum test programme will include:

- Tensile testing of skin laminates for hull panel
- Shear testing of core with skin laminates of hull panel
- Tensile testing of flange laminates for web frames, girders, beams, etc.

Until recently, little attention was paid to the need for data regarding laminate properties in more than one direction, that is to say the philosophy of the Rules was based essentially on laminates having more or less equal properties in 0° and 90° directions. However, modern practices with more highly optimised, directional use of reinforcement have led to a need for testing of laminates in two or more directions.

12.4.2 Structural Calculation Methods; Allowable Stresses

To determine stresses and deflections in GRP single skin and sandwich structures either direct calculations using the full stiffness and strength properties of the laminate in all directions or simplified methods presented in the Rules may be used. The simplified methods can be used only if the following conditions are satisfied:

- The principal directions of the skin reinforcement are parallel to the panel edges.

- The laminate elastic moduli in the two principal directions do not differ by more than 20%.
- In the case of sandwich panels, the skin laminates are thin, i.e. $d/t > 5.77$

When direct calculations are used, the Tsai-Wu failure criterion is applied considering both first ply failure (FPF) and last ply failure (LPF). The failure strength ratio R must not be less than the following:

Structures exposed to long-term static loads:
FPF $R = 2.25$, LPF $R = 4.5$

All other cases:
FPF $R = 1.5$, LPF $R = 3.3$

12.4.3 Hull Girder Strength

Longitudinal strength generally has to be checked for the craft types and sizes indicated in section 12.2.4.1. For new designs of large and structurally complex craft (e.g. multi-hull types) a complete three-dimensional global analysis is required to permit assessment of transverse and longitudinal strength.

Buckling strength may have to be checked for deck and bottom structures.

In general the minimum allowable hull section modulus amidships is given by

$$Z = (M/\sigma) \times 10^3 \quad \text{cm}^3 \tag{12.15}$$

where M = longitudinal midship bending moment from section 12.2.4.1 (kNm)

$\sigma = 0.3\ \sigma_{nu}$ N/mm^2 in general
$0.27\ \sigma_{nu}$ N/mm^2 for hydrofoil on foils
$0.24\ \sigma_{nu}$ N/mm^2 in slowed-down condition for planing craft

For monohulls, catamarans and side-wall craft M is taken from the crest and hollow landing conditions. For hydrofoils on foils M is the maximum total moment. For cases where a slowed-down condition has to be considered, M is the maximum combined still water moment and wave bending moment.

12.4.4 Sandwich Panels

12.4.4.1 General

Design of sandwich panels in twin-hull vessels is mainly determined by bending and shear strength to resist lateral loading, and follows the lines established in the classical text-book by Allen [8]. Buckling of panels has rarely been considered relevant; if such checks are to be made then recourse must be made to other works such as those by Plantema [9] and Teti [10].

12.4.4.2 Sandwich core materials

The DNV Rules specify minimum shear and compressive strengths for core materials of structural sandwich panels. For hull bottoms, sides and transoms and for cargo decks, these are $0.8\,N/mm^2$ in shear and $0.9\,N/mm^2$ in compression, corresponding to densities from about $70\,kg/m^3$ upwards for typical cross-linked PVC products. Elsewhere the minimum values are $0.5\,N/mm^2$ in shear and $0.6\,N/mm^2$ in compression.

12.4.4.3 Sandwich skin laminates

The reinforcement of skin laminates is to contain at least 40% continuous fibres.

The thickness of skin laminates on structural sandwich panels is in general not to be less than

$$t = (t_0 + kL) / \sqrt{f} \quad \text{mm} \qquad (12.16)$$

where $f = \sigma_{nu}/160$ and σ_{nu} = ultimate tensile stress in N/mm^2.

In the above, k and t_0 are specified according to location and whether the laminate is exposed or protected.

12.4.4.4 Strength and stiffness requirements for sandwich panels

The requirements for sandwich panels consist of both strength and stiffness limits, expressed in terms of allowable stresses and deflections, table 12.3. In the table,

σ_{nu} = ultimate tensile stress for skin laminates subjected to tensile stresses

= the lesser of the ultimate compressive stress and the critical local buckling stress (see below) for skin laminates subjected to compressive stresses

τ_u = ultimate shear stress of sandwich core material

σ_n = calculated normal stress in laminate
τ_c = calculated shear stress in core
w = calculated deflection at centre of panel
b = shorter side of panel

Table 12.3. Allowable stresses and deflections for sandwich panels.

Structural member	σ_n	τ_c	w/b
Bottom panels exposed to slamming	$0.3\ \sigma_{nu}$	$0.35\ \tau_u$ [1]	0.01
All structures exposed to long-term static loads	$0.2\ \sigma_{nu}$	$0.15\ \tau_u$	0.005
All other structures	$0.3\ \sigma_{nu}$	$0.4\ \tau_u$	0.01

[1] This allowable stress applies to core materials with shear elongation of at least 20%. For core materials with lower fracture elongation the allowable stress will be considered individually. For materials with higher fracture elongation than 20% an increase of the allowable stress may be accepted upon special consideration.

The Rules include procedures for hand calculation of σ_n, τ_c and w for simply supported and clamped rectangular panels subjected to the limitations listed in section 12.4.3 for the application of simplified calculation methods.

The calculations allow for shear deformation, which is often significant in sandwich panels, and are summarised below.

Maximum normal stress in skin laminates at mid-point of sandwich panel subject to lateral pressure p:

$$\sigma_n = \frac{160\ p\ b^2}{W}\ C_n\ C_1\ \text{N/mm}^2 \qquad (12.17)$$

where

C_n = $C_2 + \nu C_3$ for stresses parallel to the longer edge
= $C_3 + \nu C_2$ for stresses parallel to the shorter edge
C_1 = 1.0 for simply supported edges; given by a chart as function of a/b for clamped edges
W = section modulus for sandwich panel per unit breadth in

mm^3/mm; for sandwich panels with skins of equal thickness t having centres d apart, W = dt

Maximum core shear stresses at mid-points of edges of sandwich panel subject to lateral pressure p:

$$\tau_c = \frac{0.52\ p\ b}{d} C_s \qquad N/mm^2 \qquad (12.18)$$

where

C_s = C_4 for core shear stress at mid-point of longer panel edge
C_s = C_5 for core shear stress at mid-point of shorter panel edge

Deflection at mid-point of flat sandwich panel subject to lateral pressure p:

$$w = \frac{106\ p\ b^4}{D_2} (C_6 C_8 + \rho C_7) \qquad mm \qquad (12.19)$$

where

$$D_2 \frac{E\ t\ d^2}{2(1 - \nu^2)} \qquad (12.20)$$

for skin laminates with equal thicknesses and elastic moduli

$$D_2 = \frac{E_1\ E_2\ t_1\ t_2\ d^2}{(1 - \nu^2)(E_1 t_1 + E_2 t_2)} \qquad (12.21)$$

for skin laminates with unequal thinknesses and elastic moduli

$$\rho = \frac{\pi^2\ D_2}{106\ G\ d\ b^2} \qquad (12.22)$$

C_8 = 1.0 for simply supported edges; given by a chart as function of a/b for clamped edges

The subscripts 1 and 2 on E and t refer to inner and outer skins respectively. The coefficients C_2 to C_7 are given by figure 12.7.

12.4.4.5 Local skin buckling

Local skin buckling should be checked, but is rarely a problem with

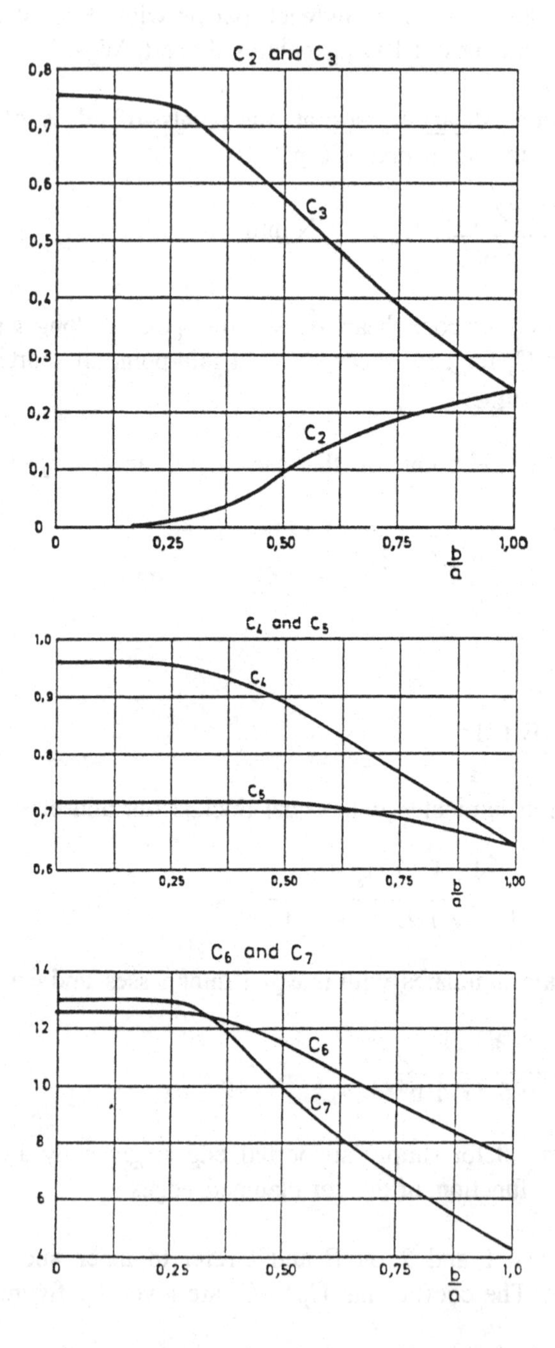

Figure 12.7. Coefficients C_2 to C_7 for calculation of stresses and deflections in sandwich panels.

larger craft. The following formula for critical local buckling stress for skin laminates subject to compression is derived in Allen's text book [8]:

$$\sigma_{cr} = 0.8 \sqrt[3]{(E E_c G_c)} \quad (12.23)$$

where

E = tensile/compressive modulus of skin
E_c = modulus of elasticity of core material
G_c = shear modulus of core material.

The modulus E_c is generally taken as the compressive modulus measured in the thickness direction.

It has been suggested, e.g. by Teti and Caprino [11], that a smaller coefficient than 0.8 should be used in the above formula. Figures in the region of 0.5 have been suggested as a result of correlation with test results. The proposed reduction is generally explained in terms of the fact that the theoretical formula does not take account of geometrical imperfections. Another contributory factor, however, is probably the fact that E_c values measured in standard compression tests are misleadingly high because the specimen experiences partial restraint in the direction transverse to the applied load.

The 1991 DNV Tentative Rules adopt the modified formula

$$\sigma_{cr} = 0.5 \sqrt[3]{(E E_c G_c)} \quad (12.24)$$

12.4.5 Single Skin Panels and Stiffeners

The requirements in the DNV Tentative Rules for High Speed and Light Craft for single skin panels and stiffeners are based on large-deflection theory that takes account of membrane stiffening. As with sandwich panels, the requirements are based essentially on the need to withstand transverse loadings; buckling related to longitudinal girder strength or local compression loads requires special treatment.

The requirements for minimum content of woven material and for minimum skin thickness are similar to those for laminates used in sandwich panels (section 12.4.4.3) but with different values of k and t_o.

The requirements for single skin panels consist of both strength and stiffness limits, expressed as allowable stresses and deflections, table 12.4.

In this table, σ_{nu} is defined as for sandwich laminates and

σ_c = combined bending and membrane stress, in N/mm²
δ = w/t,

where

w = calculated deflection at centre of panel, in mm and
t = laminate thickness, in mm.

Table 12.4. Allowable stresses and deflections for single skin panels and stiffeners.

Structural member	Laminates		Stiffeners
	σ_c	δ	σ
Bottom panels exposed to slamming	0.3 σ_{nu}	1.0	0.25 σ_{nu}
All structures exposed to long-term static loads	0.2 σ_{nu}	0.5	0.15 σ_{nu}
All other structures	0.3 σ_{nu}	0.9	0.25 σ_{nu}

The required thickness to meet the deflection requirement is expressed in terms of the allowable δ as

$$t = 178\ b \sqrt{\{p/[\delta\ E\ (C_1 + \delta^2 C_2)]\}}\ \text{mm} \tag{12.25}$$

where C_1 and C_2 are given graphically as a function of panel aspect ratio b/a, for both clamped and simply-supported edges, figure 12.8.

The combined bending and membrane stress is given by the formula

$$\sigma_c = [t/(1000\ b)]^2\ \delta\ E\ (C_1 C_3 + \delta\ C_4 C_2^{2/3})\ \text{N/mm}^2 \tag{12.26}$$

where C_1 and C_2 are as given previously and C_3 and C_4 are given similarly in terms of b/a, figure 12.9.

The formulae assume that the laminate moduli in the principal reinforcement directions do not differ by more than 20%, that the principal reinforcement directions are parallel to the panel edges, and that the pressure loading is uniformly distributed. For other cases the general requirements based on the Tsai-Wu failure criterion should be met, as in section 12.4.2.

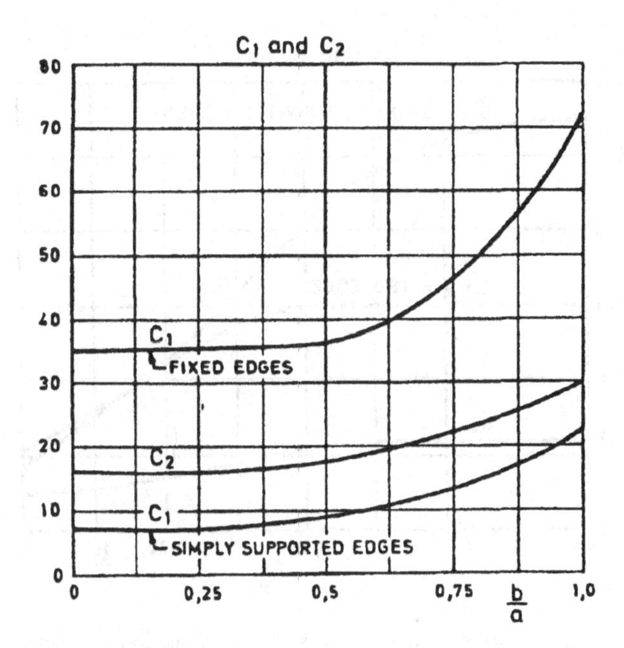

Figure 12.8. Coefficients C_1 and C_2 for calculation of thickness of single skin GRP panels.

The minimum section modulus Z of longitudinals, beams, frames and other stiffeners subject to lateral pressure is given by

$$Z = m\, l^2\, s\, p/\sigma \qquad \text{cm}^3 \qquad (12.27)$$

where l is the stiffener span in metres, s is the stiffener spacing in metres, p is the applied pressure in kN/m^2 and m is a factor which depends on the type and location of the member and lies in the range 65 to 135. The allowable stress σ is given by the values in the last column of table 12.4.

12.4.6 Web Frames and Girder Systems

For web frames and girder systems in GRP the DNV Rules give a number of formulae that can be used for the design of individual elements, based on treating the element as a beam. The requirements relate to section modulus, web area for shear, and bond area at ends of girders. The Rules require 2 or 3-dimensional structural analysis for more complex arrangements.

If the latter approach is chosen, the general requirements based on the Tsai-Wu failure criterion are to be satisfied, as in section 12.4.2.

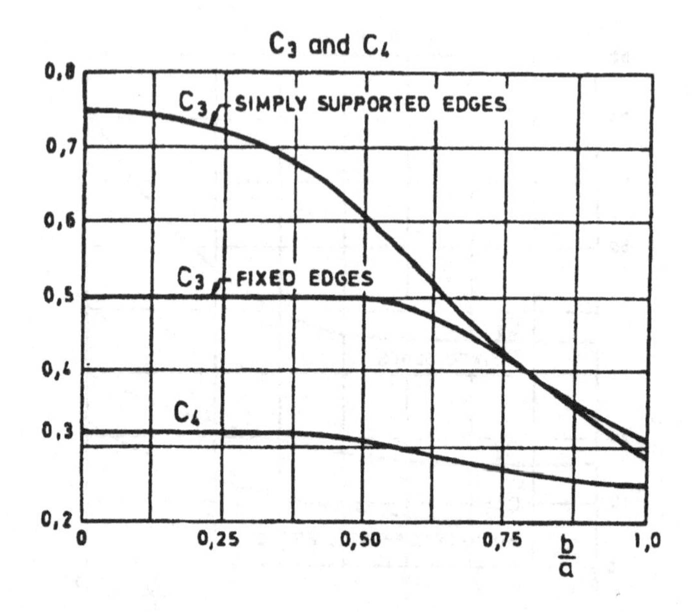

Figure 12.9. Coefficients C_3 and C_4 for calculation of stresses in single skin GRP panels.

In addition, the in-plane laminate shear stress must not exceed 0.25 τ_u.

For GRP sandwich construction the transverse frames often take a form that is more like a bulkhead with cut-outs than a slender framework, making framework analysis less applicable. Any computer analysis thus needs to take the form of a more complex finite element analysis. In practice this seems rarely to be done. The approach tends instead to be to check shear forces and bending moments at critical sections based on the global transverse shear force and bending moment as given in section 12.2.4.2.

12.4.8 Requirements for Structural Details in GRP

Requirements for continuity of structural members in GRP are similar to those for aluminium, in that well rounded brackets need to be fitted at the junctions between structural members having unequal stiffness. Brackets ending at unsupported panels need to be tapered smoothly down to zero and in addition the skin laminate on the panel needs to be locally reinforced at the end of a bracket. Girder ends are either fitted with brackets or tapered to zero, figure 12.10.

A further point requiring consideration in sandwich hulls is the danger that local damage at the bow will lead to a delamination that

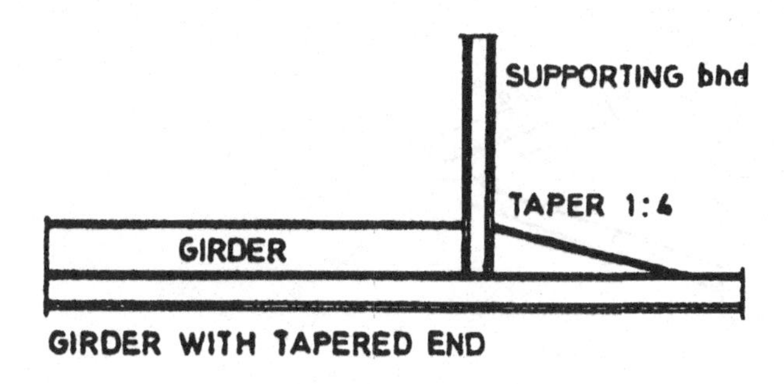

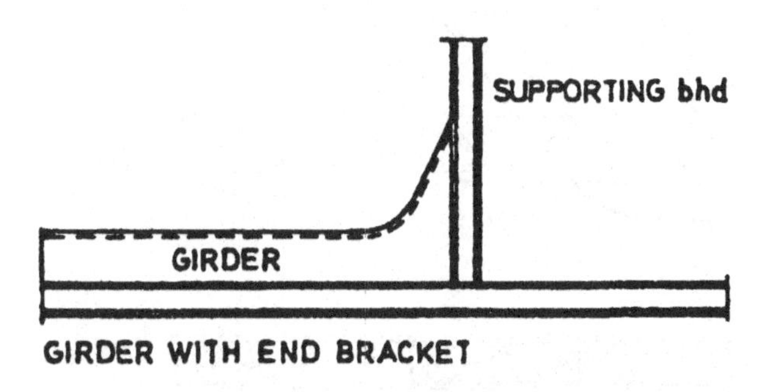

Figure 12.10. Ends of girders in GRP sandwich construction.

with time propagates further back along the hull. To avoid this, the 1991 Tentative Rules require special collision protection in the bow region for sandwich hulls. The outer and inner skin laminates are to be connected together as shown in figure 12.11, in which the distance a is not to be less than

$$a = 0.15 + 1.5\ V^2\ \Delta/10^6 \quad \text{m.} \tag{12.28}$$

The outer laminate must also be thickened in the region specified. The thickness shall not be less than

$$t_s = [7 + (0.1\ V)^{1.5}]\ /\ \sqrt{(\sigma_{nu}/160)} \quad \text{mm.} \tag{12.29}$$

In the above, Δ is the displacement in tonnes and V the maximum service speed in knots.

12.5 ACKNOWLEDGEMENTS

Contributions to this presentation have been made by several members of the staff of Det Norske Veritas Classification AS.

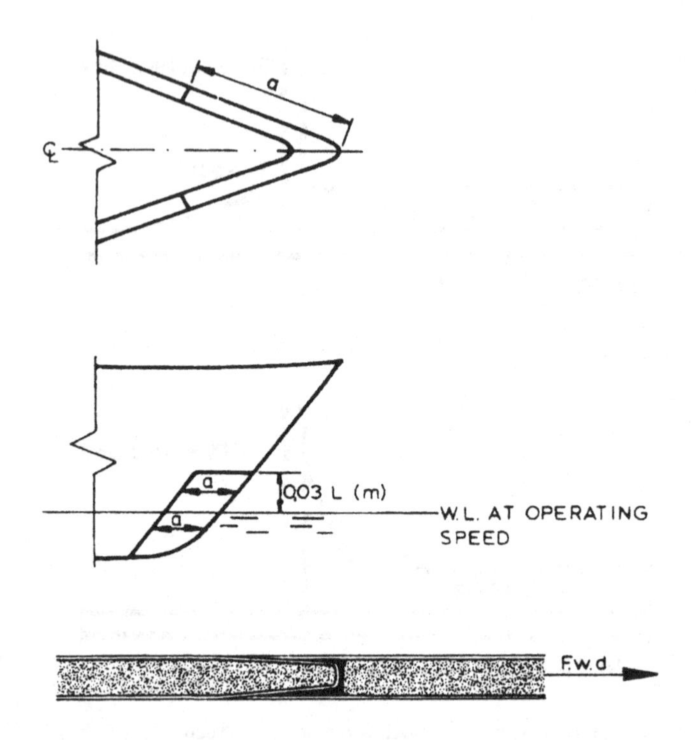

Figure 12.11. Methods of limiting effect of local damage at bow in sandwich construction. Collision protection and laminate connection.

Section 12.1 is largely based on "High Speed Light Craft Services - Introduction to Yards, Owners and Surveyors" by Karl M. Wiklund, Head of the Light Craft Department.

In preparing the remaining sections, a great deal of help was provided by Torbjørn Hertzenberg and Gunnar Haugan of the Light Craft Department.

12.6 REFERENCES

1] "Tentative Rules for Classification of High Speed and Light Craft", Det Norske Veritas, 1991.

2] "Rules and Regulations for the Classification of Light Highspeed Ships", Bureau Veritas, Paris, 1987.

3] "Rules for Building and Classing Reinforced Plastic Vessels", American Bureau of Shipping, New York, 1978.

4] Curry, R., "Preliminary Strength Standards for FRP Planing Craft", (Proposal), American Bureau of Shipping, New York, 1988.

5] "Provisional Rules for the Classification of High Speed Catamarans", Lloyd's Register of Shipping, London, 1991.
6] Savitsky, D., Brown, W.P., "Procedures for Hydrodynamic Evaluation of Planing Hulls in Smooth and Rough Water", Mar. Tech., **13**, (4), October 1976, pp 381-400.
7] Allen, R.G., Jones, R.R., "A Simplified Method for Determining Structural Design-Limit Pressures on High Performance Marine Vehicles", Proc. AIAA/SNAME Conf., *Advanced Marine Vehicle*, San Diego, 1978.
8] Allen, H.G., "Analysis and Design of Structural Sandwich Panels", Pergamon, Oxford, 1969.
9] Plantema, F.J., "Sandwich Construction", Wiley, New York, 1966.
10] Teti, C.G.R., "Sandwich Structures Handbook", Il Prato, Padua, 1989.
11] Teti, C.G.R., Caprino, G., "Mechanical Behaviour of Sandwich Construction", Proc. 1st Intl. Conf. *Sandwich Constructions*, Royal Institute of Technology, Stockholm, 1989.

13 QUALITY AND SAFETY ASSESSMENT

13.1 INTRODUCTION

The manufacture of the hulls of ships and boats or components such as piping for use in offshore units requires adherence to high levels of quality with regard to certification of processes and qualification of operators. This is very important especially in view of the fact that specification of quality through non-destructive testing and inspection of the final product is difficult to achieve.

This Chapter aims at outlining briefly the role of the Quality Assurance tool and the manner in which it can be used to control more efficiently the whole process of design, manufacture and operation of marine structures made from FRP materials, based on applicable rules and regulations. The 'input' to be taken into consideration from the preliminary stages onwards will be mentioned. At each of the design, manufacture, installation and in-service stages, particular attention will be devoted to statutory and rule requirements.

13.2 GENERAL ASPECTS OF QUALITY ASSURANCE

13.2.1 Some Definitions

- Quality: the totality of characteristics of an entity that bears on its ability to satisfy stated and implied needs (ISO/DIS 8402-1991, art.2.1)
- Quality assurance: all the planned and systematic activities implemented within the quality system, and demonstrated as needed, to provide adequate confidence that an entity will fulfil the requirements for quality (ISO/DIS 8402-1991, art.3.5)
- Quality system: the organisational structure, responsibilities, procedures, processes and resources needed to implement quality management (ISO/DIS 8402-1991, art.3.6)

13.2.2 The ISO Standards

Depending on the extent of its scope of application, ISO defines three possible standards to which the quality system has to comply:

- ISO 9001: 1987 - Quality systems - Model for quality assurance in design/development, production, installation and servicing
- ISO 9002: 1987 - Quality systems - Model for quality assurance in production and installation
- ISO 9003: 1987 - Quality systems - Model for quality assurance in final inspection and test

These three are also published in an identical form, as European standards EN 29001, EN 29002, and EN 29003 respectively. They are now universally recognised, and increasingly applied worldwide.

Guidance for selecting the most suitable standard, depending on the product, is given in the document ISO 9000: 1987 - Quality management and quality assurance standards - Guidelines for selection and use. General considerations for implementing a quality system in a firm, however, are given in the document ISO 9004: 1987 - Quality Management and quality system elements - Guidelines.

The recommended selection procedure and factors (ISO 9000) are as follows:

a) functional or organisational requirements
   - ISO 9001: for use when conformance to specified requirements is to be assured by the supplier during several stages which may include design/development, production, installation and servicing
   - ISO 9002: for use when conformance to specified requirements is to be assured by the supplier during production and installation
   - ISO 9003: for use when conformance to specified requirements is to be assured by the supplier solely at final inspection and test
b) the other six fundamental selection factors
   i) design-process complexity: this factor deals with difficulty of designing the product or service if such product or service has yet to be designed
   ii) design maturity: this factor deals with the extent to which the total design is known and proven, either by performance testing or field experience
   iii) Production-process complexity: this factor deals with:
      - the availability of proven production processes
      - the need for development of new processes

- the number and variety of processes required
- the impact of the processes on the performance of the product or service

iv) Product or service characteristics: this factor deals with the complexity of the product or service, the number of interrelated characteristics, and the criticality of each characteristic for performance

v) Product or service safety: this factor deals with the risk of the occurrence of failure and the consequences of such failure

vi) Economics: this factor deals with the economic costs, to both the supplier and the purchaser, of the preceding factors weighed against costs due to non-conformities in the product or service

In the concerned field - manufacture of FRP structures or equipment - one of the essential factors governing the choice of the model for quality assurance is the need for "special processes" which are highlighted by ISO as follows (ISO 9004, art.11.4):

"Special processes
Special consideration should be given to production processes in which control is particularly important to product quality. Such special consideration may be required for product characteristics that are not easily or economically measured, for special skills required in their operation or maintenance, or for a product or process the results of which cannot be fully verified by subsequent inspection and test. More frequent verification of special processes should be made. Keep a check on

a) the accuracy and variability of equipment used to make or measure product, including settings and adjustments
b) the skill, capability and knowledge of operators to meet quality requirements
c) special environments, time, temperature or other factors affecting quality
d) certification records maintained for personnel, processes and equipment, as appropriate."

Therefore, where special processes are to be undertaken, the criteria of ISO 9002 (or ISO 9001) are to be fulfilled, although in the reverse case, the quality system may comply only with ISO 9003 criteria. This is highlighted for the products which are mass-produced.

ISO 9001, as opposed to ISO 9002, applies mainly where the complexity of the product design justifies the coverage of the design stage within the quality system. In most cases, full compliance with ISO 9001 represents a natural evolution from a prior compliance with ISO 9002. The typical structure of a documented quality system with the cross-reference list between ISO 9004 and ISO 9001/9002/9003 is given in the figure below. It may be seen in this comparative figure 13.1 that ISO 9003 standard is less stringent than ISO 9002, which, in turn, is less stringent than ISO 9001.

13.2.3 Description of the Quality Assurance System

The main document here is the Quality Assurance Manual (QAM). Its primary purpose is to provide adequate description of the quality assurance system, while serving as a permanent reference for the implementation and maintenance of that system. It has to be set up, as far as practicable, in accordance with the framework of the ISO 9000 standard architecture.

The QAM is to be complemented by quality procedures which could be collected together in a 'Manual of General Quality Procedures'.

The quality assurance system is also to be documented by a series of documents such as job instructions, quality record forms, check lists, etc.

13.2.4 The Affected Parties

Those affected by a shipyard's or manufacturer's quality assurance system are:

- the individual departments within the shipyard (design, purchasing, production, quality control, in-coming product inspection, etc.)
- the purchaser
- the sub-contractors, including the raw material suppliers
- the classification society and other possible third parties

13.2.5 Certification/Accreditation

The shipyard or manufacturer may wish to be granted with a Certificate of Compliance with the ISO 9000 series standards issued by an accredited Certification body. The accreditation scheme is a

Clause (or sub-clause) No. in ISO 9004	Title	Corresponding clause for (sub-clause) Nos. in		
		ISO 9001	ISO 9002	ISO 9003
4	Management responsibility	4.1 ●	4.1 ◐	4.1 O
5	Quality system principles	4.2 ●	4.2 ●	4.2 ◐
5.4	Auditing the quality system (internal)	4.17 ●	4.16 ◐	-
6	Economics - Quality-related cost considerations	-	-	-
7	Quality in marketing (Contract review)	4.3 ●	4.3 ●	-
8	Quality in specification and design (Design control)	4.4 ●	-	-
9	Quality in procurement (Purchasing)	4.6 ●	4.5 ●	-
10	Quality in production (Process control)	4.9 ●	4.8 ●	-
11	Control of production	4.9 ●	4.8 ●	-
11.2	Material control and traceability (Product identification and traceability)	4.8 ●	4.7 ●	4.4 ◐
11.7	Control of verification status (inspection, measuring and test equipment)	4.12 ●	4.11 ●	4.7 ◐
12	Product verification (inspection and testing)	4.10 ●	4.9 ●	4.5 ◐
13	Control of measuring and test equipment (inspection, measuring and test equipment)	4.11 ●	4.10 ●	4.6 ◐
14	Nonconformity (Control of nonconforming product)	4.13 ●	4.12 ●	4.8 ◐
15	Corrective action	4.14 ●	4.13 ●	-
16	Handling and post-production functions (Handling, storage packaging and delivery)	4.15 ●	4.14 ●	4.9 ◐
16.2	After-sales servicing	4.19 ●	-	-
17	Quality documentation and records (Document control)	4.5 ●	4.4 ●	4.3 ◐
17.3	Quality records	4.16 ●	4.15 ●	4.10 ◐
18	Personnal (Training)	4.18 ●	4.17 ◐	4.11 O
19	Product safety and liability	-	-	-
20	Use of statistical methods (Statistical techniques)	4.20 ●	4.18 ●	4.12 ◐
-	Purchaser supplied product	4.7 ●	4.6 ●	-

Key

Full requirement
◐ Less stringent than ISO 9001
O Less stringent than ISO 9002
\- Element not present

Figure 13.1. Cross-reference list of quality system elements.

national scheme (NACCB in the UK, AFAQ in France, etc.) and the certification process is subject to strict criteria and procedures.

Increasingly, within in the European Community, the following standards are to be fulfilled:

- EN 45011 - General criteria for certification bodies operating product certification

- EN 45012 - General criteria for certification bodies operating quality system certification
- EN 45013 - General criteria for certification bodies operating certification of personnel, and
- EN 45004 - General criteria for the operation of bodies performing inspection

13.2.6 The Applicable Rules

This Chapter is primarily related to boat or ship FRP structures, i.e. pleasure boats, yachts, lifeboats, rescue boats, fishing vessels, light craft, etc. Consideration is also given to FRP piping which is used more and more on board ships and offshore units.

The main rules for such applications are published by the classification societies. For pleasure boats, there are also some statutory regulations published by national authorities such as the Scandinavian authorities. There is a European Community Council Directive draft dealing with pleasure boats; however, for the time being, there is no specific requirement related to the boat structure. With regard to FRP piping, reference may be made to the IMO document DE 35/WP7 which is a draft set up by an ad hoc working group.

13.3 DESIGN STAGE

13.3.1 Two Main Possible Design Approaches

It is necessary that the designer is able to realise the full potential from the large choice of materials to optimise the design of platings and stiffeners. This is why the methods to calculate scantlings, given by classification societies, take into account this large range of materials. Consequently, special sets of rules, quite different from those for metallic materials, have been developed.

There are two types of methods:

- the first includes those giving the weight or thickness values of laminates. They are based on 'standard laminates', where corrections are to be made when the design laminate does not correspond exactly with the "standard laminate"
- the second gives the ultimate strengths for laminates and stiffeners taking directly into account the real characteristics of the materials used

The latter is to be chosen for a more efficient design. In order to illustrate the reasons for developing special methods for scantlings determination, i.e. different from those for metallic materials, it is interesting to compare briefly the properties of metallic materials and FRP materials.

13.3.2 FRP Characteristics Compared to Metallic Materials

13.3.2.1 Young modulus and breaking strength

An examination of the stress-strain curves for standard steel, aluminium alloy and FRP is shown in figure 13.2.

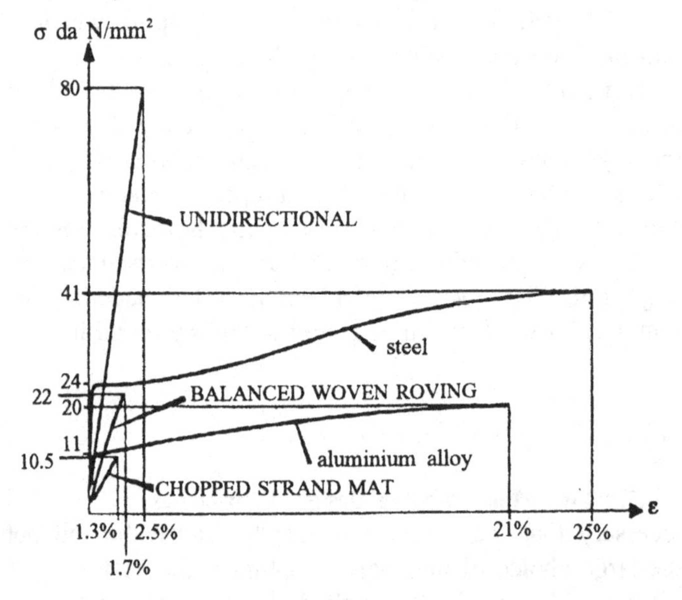

Figure 13.2. Stress-strain curves for steel, aluminium alloy and composite.

The Young modulus for standard steel is 210 GPa, for aluminium alloy 70 GPa while for GRP laminates, it depends on the type of reinforcement and its percentage in weight. The mean values are 8.5 GPa for chopped strand mat, 16 GPa for balanced woven roving and 33 GPa for unidirectional laminates. Consequently, GRP can be deformed between 6 and 25 times more than standard steel.

The yield stress is valid only for metallic materials: the values are 240 MPa for standard steel and about 110 MPa for welded aluminium alloy. For the GRP materials, there is no yield stress. The stress-strain curve is quite straight up to the breaking strength. Table 13.1 establishes the comparisons.

Table 13.1. Comparison of different materials.

	Standard steel	Aluminium alloy	GRP laminate: Chopped strand mat	GRP laminate: Balanced woven roving	GRP laminate: Unidirectional
Young modulus (MPa)	210,000	70,000	8,500	16,000	33,000
Plasticity	yes	yes	no	no	no
Yield stress (MPa)	240	110	-	-	-
Breaking strength (MPa)	410	200	105	220	800
Strain at yield stress	0.11%	0.16%	-	-	-
Breaking strain	25 %	21 %	1.3 %	1.7 %	2.5 %

13.3.2.2 Plasticity

The most important difference between metallic materials and FRP is that metallic materials used in the naval field can be defined with plasticity, while this is not the case for FRP. At yield stress, the strain for steel is 0.11% and breaking occurs at a strain of 25%, i.e. 230 times more than for yield stress. When FRP is subjected to loads, the strain-stress curve is straight up to the break.

This is of the utmost importance for scantlings determination. A metal structure is designed with a safety factor of 1.5 with respect to yield stress whereas an FRP laminate is designed with a safety factor of 5 to 10 with respect to the breaking strength.

13.3.2.3 Structure

Metals are homogeneous and isotropic. FRP is heterogeneous and anisotropic. Consequently, design with FRP is more difficult because the breaking strength is not the same in the various layers and in all the directions. Besides, there are different types of local failures such as breaking of fibres, of matrix, of interface between fibres and matrix and a combination of these modes. Breaking does not always occur at the external layer; some delaminations occur between layers as well. These points serve to illustrate the fact that the methods used for design of FRP are to be more sophisticated than those for metallic materials.

13.3.2.4 Fatigue behaviour

The fatigue behaviour of materials in general leads to a lower breaking strength, in comparison with the static case, due to a progressive degradation with the number of load cycles. Here again,

FRP materials do not have the same behaviour as that of metals. For FRP, some micro-fissures appear. Their number rises the number of load cycles, and the value of breaking strength decreases consequently.

Except at the break point of materials, no large fissures can be seen in the structure. The advantage, compared with metals, is that the FRP materials are heterogeneous and more energy is needed to break the matrix/resin connection.

13.4 MANUFACTURE STAGE

13.4.1 Fire/Explosion and Health Hazards

13.4.1.1 Fire and explosion hazards

A) Handling of raw materials

Most polyester resins are capable of burning, giving rise to flammable, volatile materials and acrid smoke, the quantity and characteristics of which depend on the composition of the resin involved.

Polyester solutions give off highly flammable vapours of the monomers in which they are dissolved, accelerators are also supplied as solutions in flammable solvents. The three substances most commonly encountered are: Styrene, Methyl Methacrylate and White Spirit. Details of their fire and explosion hazards are easily obtainable from suppliers or regulatory publications.

Static electricity may be generated when polyester resins are conveyed through pipelines to storage tanks, vessels or bulk carriers. Because of the attendant fire risk, splash filling should be avoided, and pipelines and vessels must be earthed when charging or discharging. A powder or carbon dioxide type fire extinguisher should be provided in the vicinity when discharging or charging vessels. A fire alarm point, clearly marked, should also be provided.

Lighting, heating and other electrical equipment where a flammable atmosphere could be formed must be in accordance with the national standards related to safety of electrical equipment.

All containers must be appropriately marked, and during transportation be marked, designated and documented so that sufficient information regarding their nature, hazards and emergency procedures is always available.

All polyester resins and solutions should be handled and used in well-ventilated, flameproof areas, preferably in enclosed systems. Smoking must be prohibited. If fumes or dust are likely to escape, an adequate ventilation or exhaust system is to be installed.

Under no circumstances must accelerators be mixed directly with organic peroxide catalysts; *this will cause an explosion.*

B) Storage of raw materials

Polyester resins should be stored at temperatures below 20°C, in closed containers, away from all sources of ignition, in a well-ventilated, flameproof area.

The use of stocks in strict rotation is good stores practice and helps to avoid storage times longer than the manufacturer recommends; it should also eliminate the possibility of premature polymerisation. Storage areas must be selected to avoid accidental exposure of bulk resin to fluorescent lighting, sunlight or heat.

A 'fool-proof' system of segregation for the storage and use of accelerators and catalysts must be operated: the recommended practice is to store, as far as practicable, the organic peroxides in a dedicated bunker or underground space.

C) Emergency actions

i) Firefighting: Polyesters and accelerators, particularly those containing monomers and flammable solvents, burn fiercely once ignited, giving off acrid smoke; this also applies to grades of reduced fire risk and finished products based on them. Because of the smoke and the noxious or toxic fumes produced, self-contained breathing apparatus is necessary when fighting fires in a confined space. Providing there is no danger to personnel, small indoor fires may be controlled by the use of sand, carbon dioxide, dry powder or foam extinguishers; foam is recommended in the open.

Water is unsuitable, but may be used to cool tanks, containers and drums in the proximity of fires, to prevent overheating and consequential explosion or spread of flames. Polyester resins are not water soluble. Dry powder is the most suitable medium for fighting accelerator fires.

ii) Waste-disposal: All resins and their additives are, in most countries, notifiable to the relevant authorities under the control of anti-pollution laws. Liquid resins should be absorbed on to dry sand or similar inert material and disposed of by means of approved landfill or controlled incineration.

Emptied drums, etc., contain vapours of the monomer present in the original resin solution and therefore entail fire, explosion and noxious vapour risks. They should only be disposed of by methods that follow recognised safe procedures. Operators dealing with large spillages of resin or solution should wear the appropriate protective equipment.

13.4.1.2 Health Hazards

A) General: Good industrial and personal hygiene should be observed. Personnel should be told of the nature of the products being handled.

B) Inhalation: The monomers used in polyester solutions and the solvents in accelerator solutions give off vapours that can prove harmful. In general, inhalation of vapour can cause dizziness, headache, nausea, vertigo, irritation of mucosae and unconsciousness at high concentrations. Excessive inhalation can be avoided by adequate air extraction or ventilation facilities, vapour concentrations must not exceed the relevant Threshold Limit Values.

Discharge from all ventilation and exhaust systems should be carried well clear of the workplace and of any neighbouring houses or other buildings; dust filters or arresters may be necessary with powder processes. In Scandinavian countries, the so-called 'low-styrene - emission' resins are the only ones authorised for spray moulding.

C) Ingestion: The effects of ingestion are similar to inhalation. Most commonly used monomers are harmful and cause severe irritation of the alimentary tract.

D) Skin contact: Prolonged or frequent contact with monomers or solvents removes the natural protective oils from the skin, exposes it to bacterial attack and may lead to dermatitis. Polyester resin solutions are therefore moderate to severe skin irritants. The use of a barrier cream on exposed skin is recommended to minimise dermatitis effect. Wash with a proprietary cleansing agent; do not use solvents. Gloves, overalls, and safety boots should be worn. Overalls should be laundered frequently. Heavy perspiration induced by impervious gloves can cause skin infection; it is generally advisable to wear cotton inner gloves as well and to have them washed frequently to maintain absorbency and suppleness.

E) Eye contact: Monomer and solvent vapours irritate the eyes; resin solutions cause acute eye irritation and goggles should therefore be worn where there is danger of splashing. In case of accidental eye contact, immediate thorough and continual flushing of the eye with clean water for at least 15 minutes is essential.

13.4.2 Rules and Regulations: Product Requirements

13.4.2.1 The extent of the requirements

In general, all the rules and regulations provide requirements for:

- quality control, including workshop premises
- compliance with the approved prototype
- testing and inspection at the final stage

The Classification Societies rules provide, in addition, requirements for:

- type-approval and possibly inspection of incoming raw materials
- qualification of special processes
- certification (or supervision of qualification) of operators

For instance, the European Community draft council directive related to pleasure boats provides the following production 'modules':

* module A (internal production control) for boats of less than 6 m in length and a total production of 20 boats or less per year
* module Aa (internal production control, plus tests) for boats of less than 6 m in length and a total production of more than 20 boats per year and for all boats from 6 to 12 m in length
* module D (production quality assurance) for all boats from 12 to 24 m in length

13.4.2.2 Typical statutory requirements for FRP boat structure
Hereafter references are with regard to the Scandinavian requirements from the 'Nordic Boat Standard' and/or the Bureau Veritas (BV) requirements.

A) Raw materials

Assumption has been made that the manufacturer complies with the guidelines given by the raw material manufacturer concerning the various products used for the building of glass reinforced polyester. (BV: raw materials are to be homologated, i.e.: type-approved and production to be supervised for compliance with the approved type).

B) Workshop approval

- The air temperature in the moulding premises shall not be lower than 18°C and the temperature during the moulding process shall not vary by more than 6°C. (BV: air temperature between 15°C and 25°C, variation less than 3°C).
- Hygrometry control - (BV): relative humidity not more than 70%.
- Light (UV) hazard - (BV): to take care from any detrimental effect of light on resin curing.
- cleanliness, ventilation, segregation, handling means, etc.: as per state of the art, health and safety regulations, and possible raw material manufacturer's recommendations.

C) Hand lay-up

i) Moulds (BV):

- The moulds are to be of strong construction and properly strengthened in order to keep their shape during moulding and setting of the resin.
- They are to be cleaned and dried at the temperature of the workshop before laying the release agent.
- The release agent is to be selected according to the mould's nature and the resin used for the surface layer in order not to affect its curing.

ii) Laminating by projection (BV):

- The use of simultaneous projection of resin and of fibres is to be limited to the structure areas where the access of the

projection is liable to ensure correct laminating.
- Before use, the projection equipment is to be gauged to give the percentage of fibre content. The equipment setting is to be periodically verified during operation in order to maintain the composition of laminates and the length of fibres.

iii) Surface coating (BV):
- Surface coating is to be evenly spread out with a thickness from about 0.4 mm to 0.6 mm.
- The first reinforcement is laid shortly after application of a gel coat on the mould.
- This reinforcement is to be flexible enough to ensure a good laying; it may be necessary to make special arrangements in areas where the shapes do not allow the reinforcement to correctly follow their outline.
- Upon release from the mould, the surface coating is to be examined and is not to show faults such as bubbles, blisters, pinholes and wrinkles.
- For construction moulded on inner mould, the outer surface of the hull is to be covered with a thick layer of resin or a resin-based product before painting. The resin used is to offer the properties of a surface coating.

iv) Surface coating (additional Nordic requirement):
- Emulsion bound mats shall not be used in connection with isophthalic resins.
- A light mat of maximum 450 g/m^2 on surfaces with sharp curvature and maximum 600 g/m^2 on plane surfaces shall normally be applied against the gel coat.
- The reinforcements of the laminate shall be laid in the approved sequence.
- A suitable topcoat shall be applied on the inside of the laminate in the keel and in bilge wells where it can be assumed that water can be accommodated.
- Where the laminate is not covered with topcoat or the like, the last layer of polyester shall contain wax so that the curing against air will be satisfactory.
- The overlap of two layers of reinforcement material shall be at least 50 mm.
- Polyester resin shall be applied uniformly on each layer of reinforcement.

- For at least every second reinforcement layer, the laminate shall be rolled so that the polyester becomes uniformly distributed and the laminate is as free as practicable from gas and air pockets.
- All fibres shall be well wetted but there shall be no surplus of polyester on the surface.
- The time interval between each layer of reinforcement shall be adapted to the ongoing curing process. Lamination shall not be continued on a previous layer which develops exothermic heat during curing. The time between each layer in a laminate shall, on the other hand, not be so long that the previous layer is fully cured. If it has been fully cured, the requirements for secondary bonding shall be complied with.
- During rolling over sharp edges, corners, etc., it shall be ensured that the amount and thickness of the reinforcement will not be less than required.
- Core material made of stiff foam or plywood shall, if necessary, be loaded so that it is pressed totally into the polyester during the curing process.
- All joints in the core material shall be filled before further lamination.
- The wet laminate in which the core material is laid shall be allowed to cure to some extent before further moulding on the core material is permitted.

v) Secondary bonding:
- If further lamination is carried out on a laminate which has cured for more than 48 hours, the laminate shall be ground so that the glass fibres are exposed in the surface.
- If there is wax on the surface on which further lamination is to take place, the laminate shall be cleaned unless it is not fully gelled and the wax can dissipate in the next layer of laminate.
- Topcoat shall always be ground away prior to further lamination.

vi) Stiffeners:
- Stiffeners shall be fixed to the laminate with a breadth of at least 20 times the thickness of the fastening.
- If spray lamination is applied, thickness measurements are to be carried out.

- Documentation shall be available showing that the reinforcement materials and the polyester which are used give the mechanical properties on which the approval is based with the glass percentage assumed.
- The glass percentage in the bottom laminate shall be calculated based upon the stated reinforcement weight and the mean thickness in question.

vii) Connections (BV):

- Mechanical connections by bolts, rivets or screws crossing through the laminate, used for fastening wooden or metallic elements or for fastening of superstructure or equipment, are to be made in order to avoid excessive local stresses in the laminate. They are, in no case, to go against the tightness of the laminate.

As a general rule, the protection of the laminate in the holes and openings made for the passage of connections is to be restored.

13.4.2.3 Production survey schemes by third parties

Classification Societies have developed criteria for approval of shipyards and manufacturers which have implemented a quality assurance system at their works. Its basic purpose is to ascertain the company's facilities to supply specific products from a specific facility. It is based on the assessment and approval of both the products being manufactured and the company's Quality System applied in producing those products. The products' assessment is made by reference to the relevant Society's Rules for Classification of Ships or Offshore Units and finalised by the granting of a Certificate of Type Approval. The assessment of the Quality System applied by the company for producing the products covered by the recognition is conducted in accordance with ISO 9000 series standards for certification of Quality System by a third party.

The approval of its works does not exempt the company from the inspection in production by the Classification Society's Surveyors. However, the company takes benefit of the approval of its Quality System since the Society's surveyor will adapt the traditional inspection scheme. For this purpose, the company has to draft, for each type of product or family of products, a Construction, Checking and Inspection plan (CCI plan) or a Manufacturing, Testing and Inspection Plan (MTI plan). This should specify the inspections and tests that will be performed by the works staff. For each activity, the Society's surveyor will indicate whether he needs to witness or whether performing periodic audits of the Quality Plan would satisfy him, as follows:

- C1: Mandatory hold point; the attendance of the Surveyor is mandatory
- C2: Hold point; as for C1, the Surveyor is called by the yard for inspection but may consider that his personal attendance and inspection is not necessary. He must then inform the yard immediately, so that the hold point may be lifted. The report issued by the yard Quality Department is to be kept available for consultation by the Surveyor.
- C3: The surveillance may only consist of the examination of applicable procedures applied by the builder, as well as thorough verification through random checking and counter-checking. If so, the Surveyor, although concerned, is not summoned.
- C4: The class is not concerned; the above described scheme is mainly related to one-off boats or ships built per unit. For mass-production, the current scheme is the supervision by periodic planned visits of the shipyard and production lines. For the sake of traceability, it is recommended that for each FRP boat or ship structure, a 'Moulding Planning Booklet' is to be set up describing all the moulding operations. Obviously, such a booklet may be substituted by a simple check list for mass-produced FRP structures.

  Each unit built within this survey scheme is granted at completion with a Certificate of Product Conformity issued by the Company and endorsed by the classification society's surveyor.

13.4.2.4 Non-destructive testing

Checking the safety of a FRP structure requires a verification of the quality of manufacturing. Such a verification may be made by visual inspection and, where relevant, non-destructive testing to verify that there are no unacceptable defects. There are a large number of methods, generally derived from the methods used for steel. However, when applied to composites, they present many limitations, as described below. The methods are presented according to the increasing difficulties of use or interpretation.

A) Optical examination: This method is the most simple. The product, with the light shining through, is submitted to a visual examination. Due to transparency needs, the test may be used only for glass-resin composite materials which are not covered with opaque gel coats and are of small thickness. In this manner, the general orientation of fibres and internal defects such as bubbles and delamination, are outlined.

B) Die-tests: This method, with its origins in metallurgy, allows the detection of opened defects near the surface, such as cracks and porosities. The penetrating product has to be selected according to its compatibility with resins in order to avoid harmful reactions.

C) X-Ray: When adapted from energy viewpoint to composite structures, this method mainly allows the detection of defects which are parallel to the X-ray beams. However, delamination cannot always be outlined.

D) Ultrasonic testing: The energy range and dynamic tuning to be adapted are to be compatible with propagation laws in composite materials. The two operating modes are reflection and transmission. The method allows the detection of a majority of defects perpendicular to the ultrasonic beam.

E) Infrared thermography: This is delicate to use and often needs to be carried out at night. This technique provides qualitative results to be confirmed by other methods (such as ultrasonic or X-rays) giving information on the dimensions.

F) Foucault currents: Inductive phenomena are at the base of this method, which is only applicable for carbon fibre reinforced composites (electric conductors). It allows defect determination in reinforcements.

G) Acoustic emission: Originally it was used for steel structures, this method has considerable potential for application in composite materials. The basis of the method is the emission of non-audible noises due to micro-fractures of fibres, cracks of resins, delamination fibre/resin plies. This method allows the localisation of acoustic emission sources by wave transit time measurements from the source to the various sensors, installed along the structure to be tested.

H) Tomography: This is a method developed for medical applications and is based on X-ray photography. It allows through a multidirectional survey in rotation through successive layers, for a picture to be restored by a dedicated computer. This is a powerful system, which has progressed from laboratory curiosity to industrial applications, but with high cost and limited to small sized elements.

I) Holography: This laser-based control method allows a 3-D picture of the product to be obtained. The comparison of two pictures of the same product placed in two different conditions points out levels of detachment, porosities and delaminations. This method is restricted to laboratory applications due to the size of the equipment required.

J) Extensometry: The advantage of this method is that it is cheaper and easier to apply than acoustic emission and other advanced techniques. It allows structural deformation and resulting strains to be monitored. However, the material's mechanical properties, such as Young's modulus, Shear modulus and Poisson's ratio, need to be known: these can be determined from laboratory tests on specimens subjected to strains expected on the "real" structure. This method is also a very good tool to survey the structure by means of gauges properly laid and connected to an alarm system which detects inadmissible stress or strains.

13.5 INSTALLATION (for FRP piping only)

13.5.1 Background

For components such as FRP piping, a bad installation onboard ship or offshore unit could jeopardise the whole system. This is why the International Maritime Organisation (IMO) and Classification Societies have provided requirements on the matter.

13.5.2 The Rule Requirements

Basically, the requirements for installation of FRP piping are as follows:

- Selection and spacing of pipe supports in shipboard systems to be determined as a function of allowable stresses and maximum deflection criteria.
- Heavy components to be independently supported.
- Provisions are to be made to allow relative movement between pipes and ship structure.
- External loads, where applicable, shall be taken into account.
- Joining techniques shall be in accordance with appropriate criteria. This implies that pipe bonding is to be witnessed by a qualified inspector who will have prepared written documentation of procedural aspects applicable to the work, according to the pipe manufacturer's recommendations.
- The bonding procedure is to be qualified by testing a typical test assembly.
- All bonds are to be identified and traceable to specific bonder(s) and inspector.
- The piping system shall be tested at completion according to an agreed programme to take into account the actual conditions.

13.6 IN-SERVICE SURVEY, MAINTENANCE AND REPAIRS

13.6.1 The Rule Requirements

A) Boat or ship structures

Once a boat, or ship, has been put into service, it is the responsibility of the owner to maintain her good condition. There are no specific statutory requirements from the national authorities. If the ship is classed, she has to fulfil the traditional in-survey requirements, which are:

- annual survey (especially for ships assigned with the service notation lightship, the outside of ship's bottom and related items are to be examined)
- periodical survey
- special survey (every five years: detailed survey of the whole ship's structure and attachments)
- occasional survey after grounding, damage and/or repairs

B) FRP pipes

The same requirements as for hull structures apply. IMO requires that 'at sea, the pipe material should be able of temporary repair by the crew, and all the necessary materials and tools kept on board'.

13.6.2 Survey and Repair Criteria

These matters have been dealt with in detail in Chapter 8 on Material Case Study and Chapter 12 on Regulatory Aspects in Design.

14 DESIGN MANAGEMENT AND ORGANISATION

14.1 INTRODUCTION

The objective of design management is to provide a product design that meets the design brief in an efficient and cost effective manner. It is important that this activity is recognised and respected at all management levels and that the associated organisation is structured such that the activity can be effective. It is an activity that can equally well be practised by a large organisation or an individual, thereby ensuring an effective usage of time and resources leading to a well designed product.

In this instance attention will be given to FRP composite materials and their marine application. Products made in these materials which are likely to form a part of a total product, i.e., the hull of a vessel, the chassis of a ROV, part of an offshore structure, etc., will be discussed. Aspects of design management will therefore be directed at the design activity of these composite products, as opposed to the whole product conception. However, it will be necessary to discuss the role of management at corporate and overall project level, in order to put into perspective the need and role of managing the design activity.

Products involving composite materials require very close management because of the wide variety of choice between materials, processes and structural options. Add to this the complexities of building marine structures, all of which will be affected by weight, cost and the environment, and the importance of careful management increases.

14.2 THE NEED FOR MANAGEMENT

Design management is now a recognised subject, but has only recently received the attention it deserves. The driving factor has been better and more competitive products. To achieve this, it is necessary to manage design to increase efficiency leading to reduced design costs, quicker development programmes and therefore reduced product costs to meet the requirements of the chosen market place.

Design is an interactive process, embracing several disciplines, talents and skills, including science, art, engineering, technology and manufacturing. No one discipline should have a nullifying effect on the other. It should in fact be the reverse, where each discipline strives to obtain the best compromise for the sake of an efficient and cost conscious product.

The function and necessity for design must be understood at all levels, from the corporate to the individual. In a larger organisation, design must be understood by:

(*a*) Senior management - the corporate level
(*b*) Project managers - the management level
(*c*) Design managers - the design level

14.3 MANAGING DESIGN AT CORPORATE LEVEL

At the corporate level, the overall planning is undertaken defining what is needed, and balancing this with costs and available resources. This level is also responsible for ensuring that the overall communication between levels is understood and actioned, together with the monitoring of the systems devised and ongoing evaluation.

A useful guide to the overall management of product design is BS7000 [1], which provides the following check list for senior management.

(*a*) Define, and periodically redefine, the corporate objectives.
(*b*) Make the objectives known and understood by all involved.
(*c*) Ensure that the chosen product plan is compatible with the corporate objectives.
(*d*) Provide resources to match the product plan.
(*e*) Ensure that the organisational policies and procedures are adequate.
(*f*) Ensure that those responsible for design have clear objectives, are personally motivated and motivate their staff.
(*g*) Ensure that achievement and expenditure are monitored against time.
(*h*) Maintain a sincere and visible commitment to high standards of product design.
(*i*) Evaluate achievements and communicate this evaluation to all concerned.

14.4 MANAGING DESIGN AT PROJECT LEVEL

Managing product design at the project management level ensures that the overall product (i.e., the ship, or the offshore facility, etc.) is properly controlled and managed. The project objectives will be defined, which in turn will define the objectives of the parts. In this case the requirements of the FRP structure within the overall project will be discussed. Project planning will be defined, along with costs, timescales and review procedures.

Again, referring to BS7000, the following check list is applicable for management of this overall project level:

(*a*) Ensure that a product is defined that will meet the corporate plan.
(*b*) Organise the preparation of the design brief, ensuring a wide enough spectrum of interests involved. Modify when necessary.
(*c*) Allocate budget and control expenditure and organise cash flow.
(*d*) Fix programmes, integrating the efforts of all functions, monitor progress and take remedial action when necessary.
(*e*) Ensure that the resources of all functions are adequate or made adequate to meet the programme.
(*f*) Ensure that the project organisation is adequate and that any variations from normal are made known.
(*g*) Control external communication.
(*h*) Keep senior management aware of achievement and expenditure against time.
(*i*) Organise the evaluation of the product and the management of the project.

14.5 MANAGING DESIGN OF FRP COMPOSITE PRODUCTS

Managing design of structures and components in composite materials requires recognition of the fact that design in these materials is a highly interactive process, caused by the effects of the process technology on the materials' characteristics. Furthermore there is a proliferation of candidate materials - fibres, resins, core materials, etc. with a large price range. Consequently, management is essential.

The overall objective is to ensure that a logical path is taken through the many influencing factors on design such that efficient use is made of manpower, resources and facilities. To do this it is necessary to have a formulated design process and a competent organisation to action the individual functions.

14.5.1 The Design Process

Whatever the resulting product, there is a logical methodology to the design process. However, composite structures have a much more interactive nature due to the influence of material selection, processing and the effect of the environment on the material's properties. An attempt to put this process into some form of flow pattern is made in figure 14.1 [2].

Here the overall process broken down into four main phases can be seen:

1) The brief
2) Preliminary design
3) Detail design
4) Manufacture

The importance of the first phase, the brief, should not be underestimated. There is little use in entering a long and complicated design phase if the overall objectives have not been set down clearly and the programme thoroughly costed.

This phase will therefore define the way ahead, allocate resources and set down the terms of reference. This should apply to any product, regardless of how small or large. For instance, if a high speed passenger ferry is being designed, then the overall weight will be critical. This will lead to the necessity for close weight control during the design programme and all members of the design team should be made aware of the fact that all weights must be reviewed. There is little point in having a low weight structure if weight is to be wasted in unnecessarily heavy deck equipment or excessive materials being used in the interior fit-out.

Likewise, a production sailing yacht will, more than likely, be driven by cost. Thus close attention will have to be given to the selection of cost effective materials and equipment.

Important headings in this first phase are therefore:

* a clear statement on what is required
* a nominated project manager
* defined design parameters
* a carefully selected design team having the necessary expertise, skills and resources

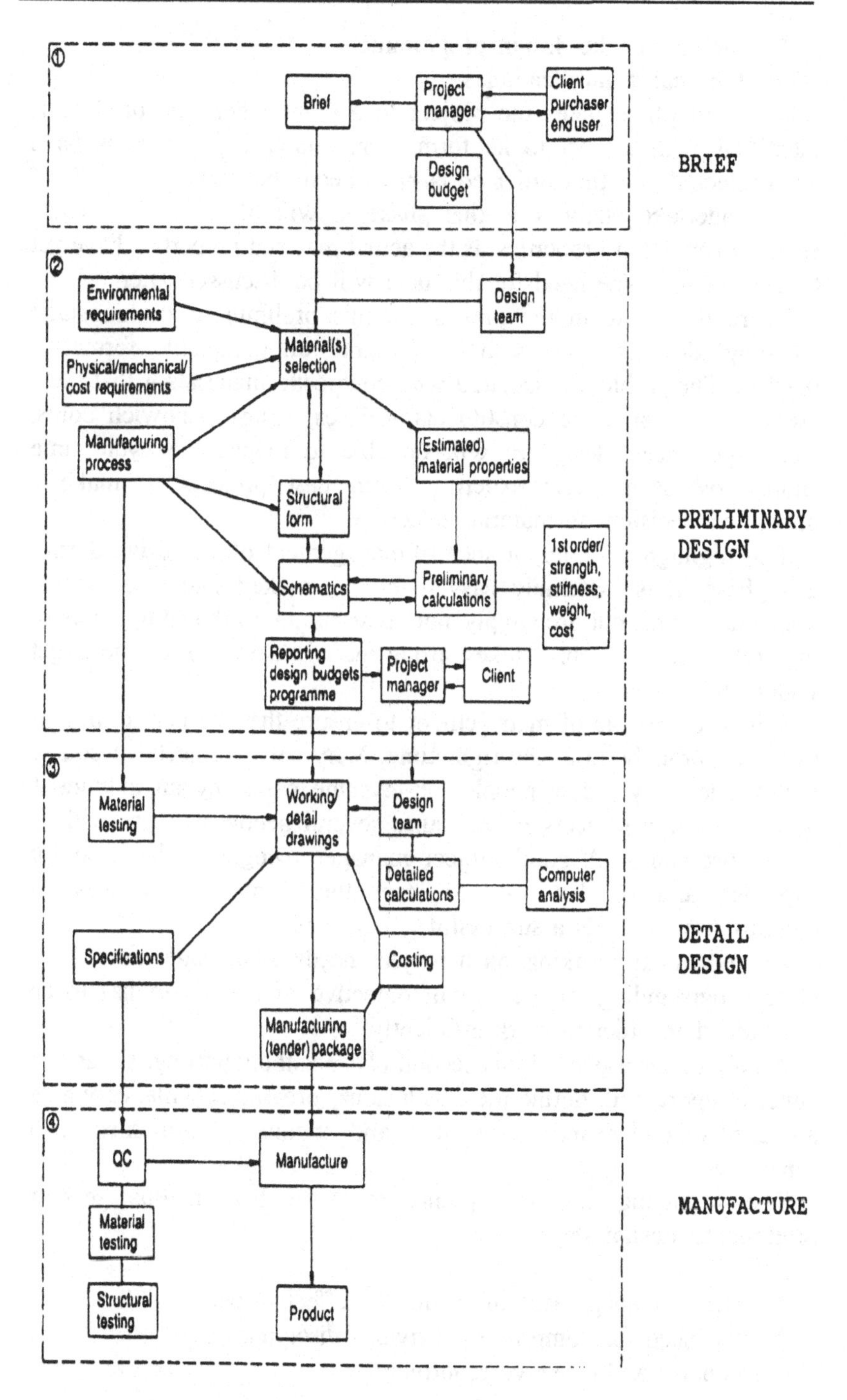

Figure 14.1. Composite structures and components.

* the cost of the design programme
* timescales and deadlines

The second phase takes the project to a point where the product is identified with respect to its format, manufacturing processes have been selected and first order costs have been obtained.

To undertake analysis in this phase, it will be necessary to use estimated material properties as the actual material properties have yet to be obtained. The need for this data will be discussed later.

There is a flow in the process in this preliminary design stage whereby ideas are put in and tried until an acceptable format is reached. The problem associated with composite materials is that there are so many candidate constituents - fibres, resins, sandwich cores, etc. Experienced designers will be able to reduce the cycle time simply by using their expertise from past projects to make a quantified decision on material selection.

This highlights the importance of management of the design during this phase. It is extremely easy to be sidetracked into areas which seem interesting but are simply not contributing to the completion of the preliminary design. It takes strong management to keep the target within the sights.

It is also the role of management to ensure that the correct design skills are brought in at the right time. Very small projects of course will be done by a few people, or in some cases, by an individual designer. Large projects may require several people as there will be many disciplines. A good project manager recognises the need for expertise and is also able to steer this expertise, by way of compromise, towards a successful conclusion.

An individual working on a project needs to manage himself. A clear understanding of the overall objective of this phase has to be maintained in order to work efficiently.

An important aspect of this second phase is the reporting. Clear and concise reports will define the conclusions, provide schematics which are easy to understand, costs, data and ongoing programmes with timescales.

The following are key points to remember in this second preliminary design stage:

* strong management to ensure the effective use of skills
* the need for compromise between disciplines
* concise and effective reporting

The third phase takes the design into detail drawings, analysis and specification stage.

The need for strong design management is essential during this phase in order to keep a close control on all aspects of the design, which invariably will include some minor and even major design changes within the phase.

There is still a flow in the process, but the flow is now directed towards the completion of a frozen idea as opposed to the creation of an idea, as in the second phase. Design management will ensure that this flow runs smoothly by allocating sufficient resources, skills and equipment.

The product of the phase will be a design package enabling an article to be manufactured by way of drawings and specifications, backed up by calculations and reports. This may be used internally for an organisation which designs and builds, or it may be used as a tender package for pricing and building by others.

The design process does not stop when the next phase, manufacture, is entered. Because of the fact that composite materials are made at the point of production, there is an ongoing flow of information back into the design loop - information on material properties, structural testing, quality control and build modifications.

The skill of design management is to ensure that the whole design process is carefully controlled through the four phases.

An ongoing activity throughout the four phases of the design process is the role that management plays in recording and controlling costs.

Large projects will have many personnel involved, whose time expended on the project must be recorded, together with the associated costs incurred - printing of documents, travel, sub-contractors, etc. The total cost of the project throughout the design phases is needed to ensure that the targeted costs are not exceeded and to have to hand valuable data when other similar projects are costed. The recording procedures must also be used to pick-up the cost of modifications and contract variations in order to negotiate additional design funds.

14.5.2 The Need for Data

The design process feeds off data which is either written down or held within an individual's expertise. Design management simply ensures that such data is available at the right time and in the right place.

The accumulation of suitable data takes time. Data for composites is aggravated by the numerous constituent materials and is not helped by the fragmented industry. For instance fibres are supplied by one company and resins by another. Data for their combined effects, i.e., a laminate, is therefore, not often provided.

If the design process is analysed, it can be seen that there are perhaps three distinct areas of data and two levels of requirements.

In the preliminary design phase 2 of the process, data is required by the design team which enables a solution to be found against the set design parameters. This data is of a reference nature and covers the behaviour of the material in the environment and in the manufacturing process, together with cost information, all of which could be summarised as "performance" data.

Also required is the material property data to allow preliminary strength and stiffness calculations to be undertaken to a first order level.

The third requirement is for the designer to have an accepted design procedure enabling the product to be designed within an internationally recognised set of requirements or standards, which will govern the design including factors of safety, loading and tolerance.

All the above data is required at two levels:

Level 1 : sufficient to generate a schematic design
Level 2 : detailed data to complete a design

A design data envelope is therefore created as shown in figure 14.2.

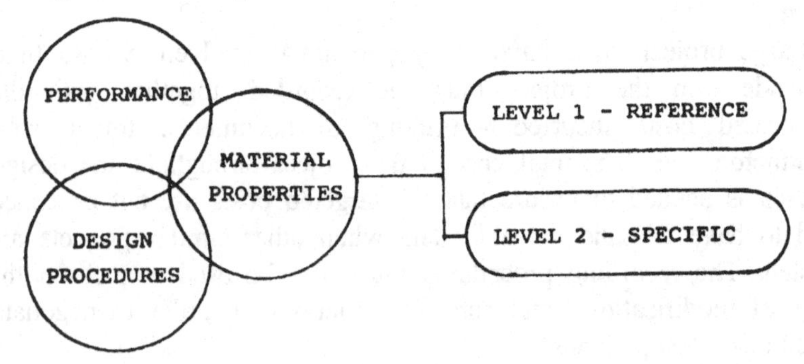

Figure 14.2. The Data Envelope.

For any product or structure, the completeness of the data is essential, it is only the degree of the content which will vary from product to product.

Phase 3 in the design process demands the Level 2 detailed and specific data to permit the design to be completed. If this phase includes the use of finite element analysis to analyse the product strength and stiffness performance to a detailed degree, then the corresponding data also needs to be detailed.

Some of the above data can be obtained from the material suppliers but definitive data has to be obtained from testing materials during and after any design and build programme. Over a period of time, a useful data bank will be formulated.

14.5.3 Design Procedures

Within the design process, there is the requirement to have recognised design procedures. This permits design to progress in a quantified manner and therefore provides easier management.

Such procedures cover the theoretical design, i.e., those factors which influence the design of the product itself and the practical side, the way in which the data is generated.

The marine industry is greatly influenced by the Classification Societies who produce "Rules" for the various types of vessels. These Rules are mainly empirical and draw upon a great deal of experience to provide scantlings for ship and boat structures, in all materials, including composites. The importance of such Rules is the set methodology which ensures a quantified means to assist design, in other words, the design procedures.

There are, however, also disadvantages. There are a number of Classification Societies, therefore there are several different sets of Rules, all of which will give different scantlings. Because the Rules are empirical, the resulting design is conservative and therefore not weight optimised. The Rules do not cover every type of vessel nor do they cover all matters of detail, where failures are often initiated.

For those projects where weight is critical, or indeed where weight is required to be saved simply from the cost aspect, it is necessary to work outside of the Rules, using first principle design and analysis.

It is then necessary to define the procedures enabling the design to progress. These will include a statement on factors of safety to be used, permissible materials, permissible methods of joining materials, imposed loads and many aspects of design detail.

This may all sound very obvious but it is surprising to find many cases where, for instance, the factors of safety to be used for a particular project or vessel have not been quantified and stated.

The other aspect of the design procedures covers the overall documentation of the design, how the many pieces of information are recorded, circulated and maintained. All this should be written down and worked-to throughout the design team. It will define, for instance, how drawings are to be numbered, who is to receive copies, how reports are to be numbered, how and where documents are to be stored, etc.

The use of CAD has greatly increased the efficiency of drawing offices and the speed with which changes can be made. More paper though is generated because any one drawing can only be seen by others when it has been plotted. This tends to increase the number of copies of drawings produced. Care is therefore needed within the design management process to ensure that this is closely controlled to avoid unnecessary data being circulated.

Another effect of computers on design has been the increase in the use of finite element analysis for structures. Because of the power of current software and the increased capacity of desk top hardware, facilities now exist to undertake major analysis with comparative ease. However, design management should address itself constantly to the need for such analysis and the usefulness of the generated data. For instance, it does not negate the need for experienced engineers in the preliminary design phase to undertake good first principle hand calculations in order to get to a working schematic as quickly as possible.

Again, returning to BS7000, the following check list for design management applies:

(*a*) Participate in the formulation of the design brief to ensure it is practical and adequately defined.
(*b*) Provide adequate design resources to meet the programme.
(*c*) Ensure that design skills are reviewed and updated by suitable training and that all design supervisors have general management training.
(*d*) Ensure that the organisation, procedures and information services are adequate and updated as required.
(*e*) Divide the tasks among the designers, ensuring that the individual and overall requirements of the brief are clear.
(*f*) Motivate all staff
(*g*) Monitor the success in meeting the brief by means of the design review and negotiate changes to the brief when necessary.
(*h*) Ensure action is taken on service experience.

(*i*) Monitor achievement and cost against time.
(*k*) Evaluate the design and quality procedures and improve these as necessary. .

14.6 INTERFACING PRODUCTION AND DESIGN

An important aspect in the design of vessels or products built in composite materials is the interaction of the design process with manufacturing aspects and processes.

Design management must recognise this particular aspect for composite materials. The material properties can be significantly affected by thc process selected. The product cost will be affected and in some cases, certain materials can only be used with certain processes.

The marine industry tends to rely heavily on the contact moulding process (wet lay-up), but more interest is being shown in resin transfer moulding (RTM) and prepreg systems. Large vessels are being built from pre-made sandwich panels, which require less tooling. Greater use is being made of pultruded sections.

Each process will have its own specific benefits which must be part of the compromise in the preliminary design phase.

A particular problem for large vessels is the question of tooling costs associated with most processes. This cost has to be amortised over the production quantity which, if too small (or indeed a single unit), may prohibit the use of composite materials. The preliminary design phase must therefore look beyond capital cost and consider through-life costs, taking advantage of the lower weight of composites and their improved environmental characteristics.

14.7 DESIGN QUALITY ASSURANCE

Design management will be concerned with establishing and maintaining quality procedures. This is no more than ensuring that the design procedures are being worked-to by means of ongoing assessment.

There are now recognised methods for setting such procedures and are laid down in BS5750 [3]. Companies who only practice in design may, however, find this document a little daunting and confusing because it applies to all types of organisations.

The basis is fairly simple though. It is principally a task of writing down what is done, the methodology of assessment procedures and then ensuring that the procedures are followed.

14.8 ORGANISATION

Design management cannot function without a specified and structured organisation. The organisation must exist corporately and at all departmental levels.

Managing the design activity will include the establishment of an effective organisational structure, recognising the following:

(*a*) The appointment of key personnel to undertake specific professional roles.
(*b*) Defining the means of communication between personnel.
(*c*) The definition of job descriptions for all staff.
(*d*) The responsibilities of sub-contracted personnel and to whom they report.
(*e*) The appointment of supporting services personnel, secretaries, administration staff, etc.

Within this organisation it is important that management continually reviews all the personnel, bringing in additional staff as required, and where necessary, additional training.

Corporate management ensures that overall company policies are maintained throughout the global organisation, the setting of staff contracts of employment and the appointment of senior management personnel.

14.9 DESIGN MANAGEMENT AND ORGANISATION - A TYPICAL PROJECT

In order to highlight the foregoing discussion, consider the design management and organisation aspects of a project where the structure of a high speed 30 metre catamaran ferry is to be designed in composite materials by a company who do not manufacture.

The brief has been written and agreed. This will include the vessel's size, speed, target weight, environmental parameters, classification requirements and manufacturing location.

As this is a high speed ferry where weight is critical, the design will be done using first principles' methods to achieve minimum weight at a defined cost.

The project manager's task is to cost out the design programme and provide timescales, having liaised with those responsible for the other aspects of the vessel's design - naval architecture, engineering, electrics, etc.

The preliminary design phase is entered with a team made up of the project manager, a structural engineer, who will also undertake any weights analysis and a designer. Each team member will be briefed as to what is required from them, the timescales, the number of hours available and the method of reporting.

The decision to use composite materials has been made for improved through-life performance, as opposed to say welded aluminium alloy, and to obtain the lightest and cheapest structure.

It is anticipated that the required quantities will be sufficient to support a female tool, thereby enabling a single skin structure to be considered as well as sandwich structure.

The design loop is now entered and various structural formats will be investigated. Many aspects will, of course, be determined by other factors, i.e., watertight bulkhead positions, overall profile and engine seats. Between these fixed structural areas, the remaining structure is investigated, looking at the various ratios of frame pitch and stiffener spacing. Each scheme will be quickly assessed for weight and cost, with many discussions with the production engineer to ensure that everything can be efficiently made.

Material data will be obtained from in-house data banks or, where no data exists, assessed by reference to relevant sources.

The skill of design management is to ensure that all the possible variations of structure can be covered without wasting time and unnecessary effort on those areas which are not critical prior to detail design. Above all, the management must ensure that minimum weight is being sought at all times within the predetermined cost constraints. This is particularly difficult for composites because of the wide range in cost of the fibres - glass through to carbon, and their effect on the weight. Naturally, the higher the mechanical properties of the laminate, the lighter the structure, but this may also mean a higher capital cost. This, however, can be offset by improved through-life costs, lower fuel costs, less power, greater range, greater speed, etc.

The manufacturing process and materials will be chosen with respect to cost, available skills, numbers required and quality, all in conjunction with the production facility.

It is only by going through the design process loop that all these factors can be assessed, eventually leading to a frozen concept.

Clearly, experienced management and designers can cut many corners because of the knowledge gained by doing similar vessels, but the process must still be followed in order to ensure an efficient, and therefore cost effective, route to an overall structural format.

Without management and a structured organisation, designers will be sidetracked, weight control will be lost, overlapping of tasks will occur and the integration with the other disciplines will not happen - all of which will lead to an unsuccessful product.

If Classification is involved, then the loads, factors of safety and material design allowable properties will have been agreed with the relevant Society at the commencement of the preliminary design phase. This will enable the drawings and design calculations to be eventually submitted to the Classification Society for approval.

On the assumption that weight and cost are within the targets set, the design moves into the detail design phase 3. The team is expanded to include further designers and engineers as required and working drawings are produced.

Each drawing will be checked for accuracy and weight, leading to a total estimated structural weight.

It is likely that test laminates will have been called for to check the theoretically derived values against those produced by the chosen method of manufacture. The results of such tests are also likely to be needed by the Classification Society for approval purposes.

The total package of information by way of drawings, specifications and calculations is completed and used for manufacture and for Classification Approval.

Modifications, that inevitably arise during manufacture, will be made, drawings and calculations altered leading to an "as built" set of drawings and calculations.

Throughout the design period, management will be controlling costs by recording time and expenses spent. Additional tasks, outside of the original brief, will be recorded for the negotiation of further funding.

Staff performance will be monitored and recorded in order to be used for future reference when other projects are undertaken.

The whole design and management process is one of actions, review and recording.

14.10 SUMMARY

Design management is necessary to ensure that a carefully controlled methodology is followed, leading to efficient use of expertise and facilities. It is particularly important for structures and components built in composite materials because of the need for an integrated approach to design and processing.

The art of design management can not be practised unless it is within a structured organisation which, not only recognises the need, but also allows it to function effectively.

Design management can be practised by individuals using self discipline to ensure efficient use of time, skills and resources.

14.11 REFERENCES

1] BS7000 : 1989, "Guide to Managing Product Design".

2] Johnson, A.F., Marchant A., "Design and Analysis of Composite Structure", in Holloway, L.C., (ed.), *Polymers and Polymeric Composites in Construction*, Thomas Telford, London, 1990.

3] BS5750 : 1987 (ISO 9001, EN29001) "Quality Systems, Part 1: Specification for Design/Development, Production, Installation and Servicing".

APPENDIX: MECHANICAL PROPERTIES OF COMPOSITE MATERIALS' CONSTITUENTS

This appendix contains information on the mechanical properties of the constituents of polymeric composites as well as properties of laminates typically used in a marine context. Information contained in the tables relates to:

1. Moduli, tensile/compressive strengths and strains of thermosetting resins;
2. Moduli and tensile strengths/strains of thermoplastic resins;
3. Moduli, tensile strengths/strains and thermal properties of some reinforcing fibres;
4. Flexural/shear moduli and tensile/compressive/shear strengths of typical laminates;
5. Shear moduli/strengths through-thickness moduli/strengths of typical sandwich core materials;
6. Fire-related properties of metalic and FRP materials.

Table A.1. Typical properties of thermosetting resins.

Material	Specific Gravity	Young's Modulus (GPa)	Poisson's Ratio	Tensile Strength (MPa)	Tensile Failure Strain (%)	Compressive Strength (MPa)	Heat Distortion Temp. (°C)
Polyester (orthophthalic)	1.23	3.2	0.36	65	2	130	65
Polyester (isophthalic)	1.21	3.6	0.36	60	2.5	130	95
Vinyl ester (Derakane 411-45)	1.12	3.4	-	83	5	120	110
Epoxy (DGEBA)	1.2	3.0	0.37	85	5	130	110
Phenolic	1.15	3.0	-	50	2	-	120

Table A.2. Typical properties of some structural thermoplastic resins.

Material	Specific gravity	Young's modulus (GPa)	Tensile yield stress (MPa)	Tensile failure strain (%)	Heat distortion temp. (°C)
ABS (acrylonitrile butadiene styrene)	1.05	3	35	50	100
PET (polyethylene terephthalate)	1.35	2.8	80	80	75
HDPE (high-density polyethylene)	0.95	1.0	30	600-1200	60
PA (polyamide, Nylon 6/6)	1.15	2.2	75	60	75
PC (polycarbonate)	1.2	2.3	60	100	130
PES (polyethersulphone)	1.35	2.8	84	60	203
PEI (polyetherimide)	1.3	3.0	105	60	200
PEEK (polyether-ether ketone)	1.3	3.7	92	50	140

Table A.3. Typical properties of some reinforcing fibres.

Type	Specific Gravity	Young's modulus (axial)* (GPa)	Poisson's ratio*	Tensile strength (GPa)	Failure strain (%)	Coeff. of expansion (axial) (x10-6/°C)	Thermal conductivity (axial) (W/m°C)
E-Glass	2.55	72	0.2	2.4	3.0	5.0	1.05
S2, R-Glass	2.50	88	0.2	3.4	3.5	5.6	-
HS Carbon (Thornel T-40)	1.74	297	-	4.1	1.4	-	-
HS Carbon (Thornel T-700)	1.81	248	-	4.5	1.8	-	-
HS Carbon (Fortafil F-5)	1.80	345	-	3.1	0.9	-0.5	140
HM Carbon (P-75S)	2.00	520	-	2.1	0.4	-1.2	150
HM Carbon (P-12S)	2.18	826	-	2.2	0.3	-	-
Aramid (Kevlar 49)	1.49	124	-	2.8	2.5	-2.0	0.04

* Glass fibres are nearly isotropic; properties of carbon and aramid fibres are strongly anisotropic and are not well defined in the literature.

Table A.4. Typical mechanical properties of FRP laminates.

Material	Fibre volume fraction V_f	Specific gravity	Young's modulus E (GPa)	Shear modulus (GPa)	Tensile strength σ_{UT} (MPa)	Compressive strength (MPa)	Shear strength (MPa)	Specific Young's modulus (E/SG)	Specific tensile strength (σ_{UT}/SG)
E-glass polyester (CSM)	0.18	1.5	8	3.0	100	140	75	5.3	67
E-glass polyester (balanced WR)	0.34	1.7	15	3.5	250	210	100	8.8	147
E-glass polyester (unidirectional)	0.43	1.8	30	3.5	750	600		16.7	417
Carbon/epoxy (high-strength balanced fabric)	0.50	1.5	55	12.0	360	300	110	37	240
Carbon/epoxy (high-strength unidirectional)	0.62	1.6	140	15.0	1500	1300		87	937
Kevlar 49/epoxy (unidirectional)	0.62	1.4	50	8.0	1600	230		36	1143

Table A.5. Typical properties of sandwich core materials.

Core Material	Specific gravity	Shear modulus		Shear Strength		Through-thickness Young's modulus		Through-thickness compressive strength	
		Absolute value (MPa)	Specific value	Absolute value (MPa)	Specific value	Absolute value (MPa)	Specific value	Absolute value (MPa)	Specific value
PVC foam	0.075	25	320	0.8	10.7	50	667	1.1	15
PVC foam	0.13	40	308	1.9	14.6	115	885	3.0	23
PVC foam	0.19	50	260	2.4	12.6	160	842	4.0	21
PU foam	0.10	10	100	0.6	6.0	39	390	1.0	10
PU foam	0.19	30	158	1.4	7.4	83	437	3.0	16
Syntactic foam	0.4	430	1070	-	-	1200	3000	10	25
Syntactic foam	0.8	1000	1250	21	26	2600	3250	45	56
End-grain balsa	0.10	110	1100	1.4	14	800	8000	6	60
End-grain balsa	0.18	300	1670	2.5	14	1400	7780	13	72
Aluminium honeycomb*	0.07	455/205	6500/2930	2.2/1.4	31/20	965	13790	3.5	50
Aluminium honeycomb*	0.13	895/365	6885/2810	4.8/3.0	37/23	2340	18000	9.8	75
GRP honeycomb*	0.08	117/52	1462/650	2.3/1.4	29/18	580	7250	5.7	71
Aramid paper (Nomex) honeycomb	0.065	53/32	815/492	1.7/1.0	26/15	193	2970	3.9	60

* Pairs of numbers refer to longitudinal and transverse directions of hexagonal honeycomb.

Table A.6. Fire-related properties of metallic and FRP materials.

Material	Melting temp. (°C)	Thermal conductivity (W/m°C)	Heat distortion temp. (°C) (BS2782)	Self-ignition temp. (°C)	Flash-ignition temp. (°C)	Oxygen index (%) (ASTM D2863)	Smoke density (DM) (ASTM E662)
Aluminium	660	240	-	-	-	-	-
Steel	1430	50	-	-	-	-	-
E-Glass	840	1.0	-	-	-	-	-
Polyester Resin	-	0.2	70	-	-	20-30	-
Phenolic Resin	-	0.2	120	-	-	35-60	-
GRP (polyester-based)	-	0.4	120	480	370	25-35	750
GRP (phenolic-based)	-	0.4	200	570	530	45-80	75

INDEX

Printed in the United States
by Bookmasters

Printed in the United States
By Bookmasters